What Is Our Life All About?

Robert Y. Ellis

First Edition

NEWMAN SPRINGS PUBLISHING
Meadville, PA

First originally published by Newman Springs Publishing 2025

ISBN 979-8-89617-599-5 (Paperback)
ISBN 979-8-89617-600-8 (Digital)

Printed in the United States of America

Prologue

Recently, I was sitting at my kitchen table on a sunny day, sipping my first cup of coffee for the day and staring out my window at the deep blue Atlantic Ocean and the bright blue sky above filled with scattered, constantly moving puffy, white clouds. I found myself filled with wonder over the beauty of this ever-changing scene, which I view most every day. There are all kinds of scientific explanations for our weather, but I still couldn't help being awe inspired by this beautiful display of nature.

I've been told by some that my head is filled with all kinds of weird wonderments. I guess it's true because I suddenly found myself wondering about our planet Earth and the part it plays in the universe in which we reside.

And I began to wonder how many folks think about the fact that we are residing on an infinitesimal speck in this universe, riding round and round a star.

Depending on where we are on this planet we call Earth, we are traveling as fast as some 1,000 miles per hour as our planet revolves around its axis. If we are on the North or South Poles, then we are spinning around as we stand still. Strangely, we don't get dizzy.

As well, our Earth travels around our sun about 584 million miles a year. So while we are revolving on our Earth, we are also traveling some 66,667 miles per hour around our star. That's 1,111 miles per minute. As I sit in my chair, typing these words, I am not aware of these manifestations of our Earth's participation in the actions of our solar system, universe, universes. Are you?

We say that we need to save the Earth, but our Earth keeps revolving around the sun and spinning around its own axis and will continue to do so until a few billion years from now when, astrophysicists tell us the sun will explode and engulf the Earth.

It would seem, then, that at least for the present, our Earth does not need saving. What needs saving is life on our Earth in all its forms. The planets will keep going round and round. The stars will keep glowing. Our galaxy and all galaxies will keep spinning.

We are here—all of us of all kinds of racial and ethnic diversities; of all kinds of sometimes extremely diverse beliefs and understandings; all of us, believers and nonbelievers in everything, from an amazing collection of religions, of atheists, agnostics, believers in science, nonbelievers in science—as long as we occupy this space on this planet, in its galaxy, in this universe. Why can't we get along with one another?

*****

In her book, *The Sixth Extinction,* Elizabeth Kolbert[1] describes the process by which our current human species are facilitating the extinction of our own existence. We humans now on this Earth are working toward becoming extinct within not that great a time, she says. There have been at least five mass extinctions in the history of life on Earth, and four in the last 350 million years. We are on the brink, says Ms. Kolbert, of bringing about our own extinction. A sixth extinction caused by huge overpopulation, colossal global warming brought on by human activity. As well, we are doing our best to extinguish many species—wolves, rhinoceros, lions, tigers, hypos, gorillas, elephants, chimpanzees, polar bears, insects, whales, dolphins—all because we humans, who are occupying more and more space on this planet, can't get along and, therefore, can't agree it's time to change our ways.

What if we, *Homo sapiens* of all kinds of colors, ethnicities, belief systems, were not here?

All other creatures, great and small, would have continued their existence subject only to the laws of nature. Life on this planet would continue.

---

[1] See https://www.pulitzer.org/winners/elizabeth-kolbert.

So why does all the pain and suffering and all the joy and happiness we experience on our planet Earth matter? Why do we, *Homo sapiens*, concern ourselves over life as it exists on our planet? Why, for instance, should I care about the possibility of the United States becoming a dictatorship?

No matter what amount of ranting and raving various *Homo sapiens* engage in, no matter what inability of said *Homo sapiens* to get along with one another, no matter how many insurrections, riots, it probably won't matter a hill of beans. The human species is likely to go extinct. Possibly other life forms—vegetable, various mammalian, aquatic, insect—may survive but not *Homo sapiens*?

However, projections are being made that a superior "intelligence" will replace human intelligence. And this superior intelligence will have, ironically, been created by human intelligence.

Well then, will this "intelligence" be a life form?

Billions of dollars are being spent in various laboratories all over the world, developing algorithms which can learn. Many relatively crude-learning algorithms exist right now. You've experienced their learning powers. They make predictions regarding how you may react to various stimulations, such as your buying habits.

Much more sophisticated algorithms are being developed, which will learn how to react to various stimuli—good or bad—regarding our human existence, in medicine, and eventually in all walks of life.

*****

What is our existence all about? Perhaps more to the point: What has human existence been all about since it began? We are known as *Homo sapiens*. *Homo* refers to man-like two-legged creatures, and *sapiens* comes from the Latin for "one who knows."

Early on, there were several humanlike creatures walking on this Earth—*Neanderthal*, *Homo erectus*, to name two. Slowly but surely, all but *Homo sapiens* were extinguished, Neanderthal being the most recent. Why did these other creatures disappear? Because they could not survive the conditions they faced. Will we be able to survive the conditions we are creating?

There is some speculation that the Neanderthals were driven into nonexistence by us *Homo sapiens* but not before there was a certain amount of intercourse between the two leaving some Neanderthal blood flowing in the veins of us, *Homo sapiens.*

It would seem, then, that it was our superior intelligence that made it possible for *Homo sapiens* to win the two-legged, mankind race. But just think of all the incredible pain and suffering that accompanied this win. How many individuals, of whatever hominid, suffered pain and suffering at the hands of their fellow hominids? *Hundreds of thousands of years of suffering.*

We don't have to look that far back to see such cruelty. Less than one hundred years ago, we witnessed an attempt by a few *Homo sapiens* to eliminate the entire population of Jews from this Earth.[2] The Nazis called it the final solution. The Nazi reign of terror was not the first. Such attempts to overcome what were thought to be undesirables have gone on since time immemorial.

Even today, we are witnessing attempts by various groups of *Homo sapiens* to reign terror over their neighbors, who they consider to be undesirable. Think of the so-called White Christian nationalists. Think of recent hateful diatribes against nonwhite-skinned immigrants.

But let's set aside the *Homo sapiens'* directed cruelty and look at the unfortunate members of our species, who simply are down and out. This is one area where we, *sapiens*, can claim some solace. There are thousands of governmental and nongovernmental organizations which are attempting to address the evils facing us *sapiens* today and bring some level of comfort to the suffering.

But I ask, Will it never end? These various genocides, holocausts, and all the poverty that exists worldwide, have always been with us. For as long as we, *Homo sapiens*, have trod this Earth these evils have gone on.

All to what end? What is the point of our hominid existence?

---

[2] For so-called Holocaust deniers, I visited Auschwitz-Birkenau in Poland. I stood in one of the gas chambers and saw the crematorium. It was an incredibly moving and troubling experience.

This question, of course, assumes that some kind of initial design started it all, had a meaningful purpose, and that it all went awry—really awry, starting at the beginning when initial life forms started, immediately consuming other life forms for the purpose of surviving and being able to procreate, which process continues to this day.

Is there no point to it all? None? Should we say that everything just is? All that munching by us of our fellow animals—cows, pigs, fish, et al.? All the munching of their fellow creatures and plants, all of hominids' destruction of the very living things upon which we live? All of it?

All the pain and suffering and angst-producing activity, simply *is*?

In his book, *Sapiens: A Brief History of Humankind*, anthropologist and historian Yuval Noah Harari[3] explains how *Homo sapiens* came to be the sole human species, describing various *Homo* species that evolved and went extinct over a period of approximately two million years.

He ends his description by questioning whether human progress has led to increased human happiness. He concludes that it is impossible to show that it has.

If that is the case, then what does the future portend?

*****

In 2005, inventor and futurist, Ray Kurzweil,[4] wrote *The Singularity Is Near: When Humans Transcend Biology*, in which he projected a future where artificial intelligence would become supremely important. Among other things, he suggested that implants for the brain would be developed, which would supersede any powers of calculations we humans now have, even more powerful than the most advanced mathematicians and scientists—thanks to AI.

In 2016, Harari published *Homo Deus: A Brief History of the Future*. Here he takes Kurzweil's predictions to a future, in which

---

3 https://en.wikipedia.org/wiki/sapiens:_A_Brief_History_of_Humankind.

4 See https://en.wikipedia.org/wiki/The_Singularity_Is_Near.

developments in medicine and AI will bring us to where we will no longer be *Homo sapiens*, rather we will become a new species, which he suggests might be called *Homo Deus*. *Homo* means "man." *Deus* means "God." His suggestion is that AI and all the physical and medical developments that will come out of it will make it possible for a new species to evolve, which will be able to live forever, not be immortal. Death caused by accidents, murder, et al. will occur.[5]

Like it or not, this life expectancy statistic is incredibly important. It is we, *Homo sapiens*, who have brought on all of our survival issues. Some of us are trying to reverse these trends. But can we reverse them in time to save ourselves?

We are unbelievably divided religiously, politically. Wars upon wars are a part of *Homo sapiens*'s existence, demonstrating the incredible pain and suffering we, *Homo sapiens*, are capable of inflicting upon one another.

All kinds of religious and political potentates will insist on inflicting their megalomaniacal beliefs on the rest of us. The list goes on and on, reaching, in the United States, right up to the United States Supreme Court. Political potentates, who care about nothing but their own comfort with their megamansions, personal jets, private airstrips, and their own egotistical sense of importance, appear to have no interest in the future other than their own.

Women in the United States are fighting fiercely to overcome the rampant onslaught, by a bunch of mostly old men in high places, along with so-called Christian nationalists, Biblical fanatics aimed at denying their right to have control of their own health care, especially abortion rights. And as if that's not enough, in the US, we have the ongoing battle over the right to bear arms with groups of people being shot dead on an almost daily basis. Then there are the more silent issues. The constant onslaught on the US Constitution's guarantee of separation of church and state and all sorts of attacks on our rights as citizens to exist without governmental interference.

---

[5] See https://www.britannica.com/topic/Nonbiological-Man-Hes-Closer-Than-You-Think-2118807.

How many megalomaniacs can we list who are infecting our way of life right now? Worldwide. There are thousands of *Homo sapiens* in government and the business world who thrive on their megalomaniacal lives. And there are the literally billions of us, *Homo sapiens*, who are lost, swimming in drugs, poverty, and sickness without aid.

Two thousand five hundred years have gone by since the birth of Christianity's Savior, whose golden rule is incredibly important to humanity's survival. And yet I have had to come to the conclusion that, despite the good intentions of so many folks that I know and others trying to alleviate the pains and fears of so many humans and wild and plant life, our survival, worldwide, seems hopeless.

At my age, I probably need not be terribly concerned. The woodlands that surround our home, the deer, raccoons, squirrels, chipmunks, foxes, coyotes, wild turkeys, and others that come by from time to time probably will continue to do so.

My wife and I will continue to enjoy the love and company of our son and daughter and their children and so many friends of all kinds of religions and beliefs, which give one a sense of hope. I will continue to support a number of charities, which support the causes I am concerned with.

As I write, I peer out my window from time to time. I see blossoms popping out of the branches of various plantings in abundance. A squirrel just ran across the deck. A while ago, a deer went by, followed by her very cute young fawn. Seagulls fly by. Various birds are singing their songs. We have several crows, who populate our trees with their loud raucous calls. Not long ago, I saw my raccoon friend in the woods.

And it all makes me feel sad. In the last analysis, for me personally, what has been the point of my ninety-one years of existence? I suppose the point has been my marrying my wife of sixty-four years; the raising of our children; their raising of their children, our grandchildren. I, personally, have been blessed.

But…

1

Our store was open from 9:00 to 6:00, Monday through Saturday. Located at 1968 Amsterdam Avenue in Manhattan, New York City, it also was a factory. When my father first moved into this location in 1934, the year I was born, it had been a movie theater. Now it was a carpet cleaning plant and a carpet, linoleum, and tile store.

The store had originally been the lobby of the theater. A long, narrow space, about fifteen feet wide and perhaps forty feet long, it was entered from the sidewalk through a door, with a floor-to-ceiling glass window. Next to it was a glass enclosed display area in which various products were shown—samples of carpets, bottles of floor shine, samples of floor tiles, rugs, scatter rugs, bottles of spot cleaners.

At about 5:00 on this particular evening in 1967, I was in the store area when a woman and her daughter came through the door. "Oh, I am so glad you are still open," the woman said. "We need some linoleum." Her eyes roamed over the rolls of linoleum that stood on end, filling most of the store.

Linoleum came in rolls we had standing on their ends. Some stood six feet tall, others nine feet. They could be very heavy and dangerous if they fell over. My father had designed a system of aisles with colored ribbons, which, when tied, held the rolls in place so they couldn't fall over accidentally. It was somewhat like a library with short aisles, seven or eight feet long, extending out from the wall down which you could walk and look at all the various patterns and colors available.

"All right," I said. Why don't you have a look. Pick out something you like, and I'll go over to your place tomorrow, measure the space, and give you a quote for the linoleum and for having it installed."

"Oh," she said, "but we need it tonight!"

"I'm so sorry," I said, "but our installers have finished for the day. They've gone home. And we need to measure the space."

"But we've measured it already. And you can install it," she said.

"But, ma'am, I've never done an installation. You wouldn't want me to ruin your room."

"But you know how, don't you?" It was more a statement, than a question.

"I've watched our installers, but that's not the same."

"Well, we know you can do it, and here are the measurements." She handed me a ball of string, which she pulled out of her purse. She undid the roll and said, "The room's this wide." She stretched the string out from a knot at the end to a knot in the middle. "And it's this long," she said as she continued unraveling the string to a second knot. This was not unusual. Many customers came in with measurements they had taken with string.

I would have to check the string with my tape measure. But as we undid the string, I could see we would need nine-foot-wide goods.

"Ma'am," I said. "Our trucks are put away. There's no way we could get the linoleum to your place."

"We can help you carry it. We're only two blocks away," she said.

Oh, boy, I thought, *they had this all figured out before they ever came in.*

I stood there with my mouth slightly open. Here were a mother and her rather pregnant daughter, now staring at me with what could only be a sense of hope in their eyes. Her daughter was very cute and had a bright smile on her face as she looked at me. Her mother, too, was smiling and nodding her head encouragingly.

If I had to guess, the mother was in her late thirties, maybe five foot two, dark skin, large, slightly slanted brown eyes as I recall. Her black hair probably had been straightened. She still had her girlish figure in spite of having born her daughter. She was a very attractive woman. Her daughter was somewhat of a copy of her mother. Though her hair was very curly. They both had engaging smiles as they looked at me.

Our store was in Harlem, located on Amsterdam Avenue between 157th and 158th streets. Our trucks went all over the city, picking up dirty rugs for cleaning and delivering cleaned rugs or installing linoleum or wall-to-wall carpet for customers who had phoned us and upon whom we had called and for whom we had taken measurements. Pretty much all our walk-in customers were from the neighborhood. Usually, a customer would come into the store, pick out a pattern of linoleum they liked, and we would go to their apartment to measure the room, often a kitchen.

My father and mother lived in Yonkers, and my wife and I lived in Riverdale in the Bronx. Except for me and my father, all the staff—secretarial, factory workers, installers—were African American. I often worked in the factory or in a truck alongside these guys. I took all my instruction from them, including, in theory, how to install linoleum.

We had a routine. We, in the store, sold the goods. Our installers, some of whom got paid according to the number of square yards of linoleum they installed, did the installation. It was not up to me to take work from them.

This woman and her daughter might be an exception, it seemed.

While I had never installed any linoleum, I had watched a number of installations. Only one tool was required, in general. That was a specially designed and very sharp knife. We never told our customers, but this kind of linoleum that we were looking at was intended for what was called a flop job. The linoleum was spread out on the floor, and the knife was used to cut it into all the various curves of the baseboards, around door jams, and other obstacles. No cement was used. The linoleum was simply flopped down.

Often the new linoleum was installed over the old one that was being replaced. That meant that the floor got higher and higher as the years went by. And that sometimes meant that the old had to be removed to make room for the new, otherwise the door might not be able to swing open.

All this information was buzzing through my head as I asked, "Will this be going over another linoleum already there?"

"No," the woman said, "we already took it away, and we've cleaned up the whole room and took out the bed and furniture. All you have to do is put it down."

All I had to do. "Are you sure this can't wait until tomorrow, when we'll have an installer able to cut your linoleum in properly?"

"No, we can't wait. It's Charlene's birthday, and we're having a party with her intended and friends later this evening. And the furniture's in the kitchen. It can't wait."

So this cute, pregnant young lady, maybe sixteen years old and with an engaging smile—really very appealing—was Charlene and was engaged to be married, and they were celebrating her birthday in a couple of hours. I looked at her and smiled. I mean, how could I refuse?

"Okay then," I said somewhat hesitantly, "let's have a look at the string." She handed it to me. I removed my tape measure from my belt and ran the string along the tape as I unrolled it from its case. The mother held onto the end of the tape and string as I played the tape out. The first knot came to eight feet and seven inches, so we would use nine-foot goods. I continued playing out the tape until we got to the second knot—just under twelve feet.

"So it looks like the room is just under nine-by-twelve feet. Does that sound right?" I asked.

"Yes," she said. "I have a yardstick and measured it with that. That's what I got."

"Okay, so you'll want to choose from these rolls, which are nine feet wide," I said. These linoleums came in various thicknesses and stiffness. If it was really stiff, I knew it would be harder for me to cut in accurately. I hoped she'd go for something softer. If it was soft enough, I could spread out the linoleum on the floor and press it in against the baseboard so that I had a crease I could reference when cutting. Stiffer stuff meant I would have to sort of guess where to cut until it fell in place. If I cut too much away in that first cut, I could have a real problem.

Mother and Charlene roamed among all the nine-foot-tall rolls. "How do you like that?" Mother would ask.

"It's too dark," might be Charlene's response.

The price for the rolls was marked on a tag that was stuck to each roll with Scotch tape. They ranged from $0.99 per square yard to $1.99. The more expensive would come to about $24. I was pretty sure this was a special and rather expensive purchase they were making. But price didn't seem to matter. They were looking for a right color and pattern.

Finally, Mother said, "We like this one." It was a terra-cotta tile pattern, predominantly in gold. *One of the nicer patterns,* I thought. And it was not too stiff!

"Okay," I said. I grabbed a sales slip and filled it out. "What's the name?" I asked.

"Harriet Brownlie. That's b-r-o-w-n-l-i-e," she spelled it out. "Everybody thinks it ends in l-y, but it's l-i-e."

"Right. And the address?"

"We're at 263 West 156th Street. We're on the second floor, so it shouldn't be too difficult to get it up there." She was smiling very broadly at me. They had picked a pattern that was $1.99 per square yard. The total square yards came to twelve, so a total of $23.88 for the linoleum and $1.19 for the tax—total $25.07. I showed her the sales slip and hoped it wasn't going to be too much for her. It wasn't. She handed me a twenty-dollar bill and $6.

"I'll be right back," I said and walked to the office at the back of the store.

"You won't believe what I've just done," I said to my father and Mildred who was his secretary.

"What's that?" they both responded.

"I just sold a nine-by-twelve piece of linoleum."

"And?" Mildred said.

"I've agreed to install it tonight."

"What?" my father said. "You mean to those two women out there?"

"Yes. That's a mother and her daughter. They're celebrating the daughter's birthday later this evening and need a new floor."

"But you've never actually done an installation, have you?" Dad asked.

"No. And I told them that. But they convinced me I would be able to. Don't ask how. And they have the room all cleared out and cleaned, ready for the linoleum. And I have helped Sammy and Eldred at various jobs."

"How are you going to get it there?"

"They're only over on 156th Street, and they are going to help me carry it."

"That should be a sight to behold. Let me go out and say hello to them."

I made the charge slip for the sale and brought it and change out to Mrs. Brownlie and Charlene. My father followed me.

"Hello, Mrs. Brownlie, I am Bob's father," he said as I handed her the change and sales slip.

She smiled. "It's nice to meet you."

"Are you certain you want Bob to install your linoleum?" he asked.

"Yes, we are. We'll help, and he'll do fine."

"All right then"—he turned to me—"I'll wait until you get back. Good luck." Then he turned to Mrs. Brownlie and Charlene. "And if there's any problem, you be sure and let us know."

"Oh, there won't be any problem," she said. "But thank you anyway."

My experience in calling on folks in this neighborhood, dealing with them in their apartments and in the store, was that they almost invariably were very nice people. *But Mrs. Brownlie and Charlene were amazing*, I thought. Of course they were asking for a lot. But still I couldn't help but feel a sense of affection for them.

Dad headed back to the office, and I said, "Okay, we need to cut twelve feet off this roll." I headed over to a small desk and grabbed a knife. "Now I am going to ask you guys to help me out a bit. I am going to pull the linoleum off this roll. I'll measure it as I go along. It will have a tendency to want to roll back up. So as I hand it to you, just let it roll up and create a new roll."

I started to pull off the linoleum.

"Aren't you going to use your tape measure?" Charlene asked.

"I don't need to. I'll show you. If you look at the pattern, you can see that it repeats. On this pattern, every sixth tile marks three feet. See?" I pulled off about four feet, and they could see what I meant. "So I need to pull off four of these repeats, right?"

"Yes, I see." Mrs. Brownlie said.

We pulled off the four repeats. I cut the linoleum from the top down to the floor, and we had a twelve-foot-long roll of linoleum, which stood nine feet tall. I grabbed some string and tied the roll, then I headed toward the back. "I'll be right back," I said.

I went to where the guys stored their tools and grabbed a belt with a tool pouch attached. In it was a knife and blades, a screwdriver, hammer, pliers, and various other tools one might need. "Okay," I said as I put the belt on and headed back to where they were standing. "I guess it's now or never."

They smiled at me as I tipped the roll over, until the upper end rested on my shoulder. I turned my back to them so that I was facing the door. "If you can pick up the other end, I'll open the door. We can pass through, and you can close the door behind you."

At that moment, my father came out. "You all go on out," he said. "I'll close the door and lock up. It's almost six. Just knock on the door when you get back. I'll be waiting for you. Good luck."

We got out on the rather broad sidewalk, and Mrs. Brownlie suggested, "Why don't we turn around? Then Charlene and I can lead the way to our apartment building."

"Okay. That's a good idea." Other people were walking back and forth. But they all stopped when they saw us with our nine-foot-long burden, and we were able to turn ourselves around. Once we were all set, a few people applauded. It must have been quite a sight for them. And off we went.

About ten minutes later, we arrived at their apartment building. The entrance was up five steps, and the door was closed. "I can carry this," Mrs. Brownlie said to Charlene. "You open the door and hold it while we go through." Then she said, "There's no elevator, but we'll be okay going round the bends. And we're only on the second floor."

Most of these buildings were walk-ups. There were no elevators. They usually were six floors high. I was glad we only had to make it to the second floor.

We made it up the first flight of stairs to a landing. The roll could not be bent. That would ruin it. So as we got to the landing, we stood the roll on end on the landing and then let it fall in the direction of the second flight. When we got to the second floor, we repeated the process until we could point the roll toward their apartment and repeat it to get the roll into their front door.

Once we got the roll into their apartment and their door closed, Mrs. Brownlie suggested we put it down on the hall floor, rest a bit, and they could show me where the room was.

The hall ran down the middle of the apartment with rooms on either side and a fairly large living room at the end, which looked out over 156th Street. Charlene led me down the hall to the second door on the right. We walked past a kitchen on our left, which was filled with furniture and a bed, which had been taken apart. I was very happy to see that, and Charlene's room was empty and very clean. Also, the hardwood floor was in good condition. Sometimes you had to look for nails sticking up, which could punch a hole through the goods, hence the hammer. But this was pristine.

I had been in many of the apartments in our neighborhood, taking measurements for both linoleum and carpet installations. This was Harlem in 1967. It was an all-Black area. I was always struck by how neat and clean the apartments I visited were, contrary to the myths propounded by us Whites.

"Okay," I said. "Now how do we get this roll into the room?"

"I think," Charlene said, "if we bring that end past the door, we can lift it up to the ceiling, and then the other end can go into the door."

"Okay. Let's try it." Charlene had been mostly quiet up until now, but I guessed she was getting more at ease with me. And after all, this was her room. Her mother stood by as Charlene and I lifted the roll off the floor, walked down to where my end was at the door, and then she lifted her end up to the ceiling and moved it over to the wall opposite her door, while I lowered my end to the floor. I had to squeeze my end a bit. But it fit through the door, and we were able to get it all the way in without damaging the goods.

We set the roll on the floor. I grabbed my knife and cut the string. Both the gals were in the room. I handed the string to Mrs. Brownlie and then grabbed the end of the goods from the middle of the roll and pulled up as far as I could. The linoleum was reasonably flexible, and I was able to pull off perhaps seven or eight feet. The balance of the roll was against the wall, opposite where I was standing. I let what I had pulled out fall to the floor, walked over to the rest of the roll, and managed to unroll it upside down back over what I had pulled off. Without saying anything, Charlene walked over to the far corner of what I had just pull out, and her mother went to the other corner. I went back to the first section I had pulled off and grabbed it in the middle of its nine feet with my arms spread out as wide as I could. Without anyone saying a word, the three of us proceeded to swivel the linoleum around so that finally I had my nine-foot-long end against the outer wall of the room, and the women were at the entrance to the room. The piece was spread out nicely on the floor. We adjusted it a bit to be certain the pattern was straight, relative to the walls.

"Okay, time for me to go to work." I had helped out on a number of installations, so I did, in theory, know the procedure. I grabbed my knife and started cutting. Fortunately, the room was reasonably square. I worked my way around the room, making shortcuts and trimming more and more away, until finally the linoleum fell to the floor against the baseboard all around. There were some trim cuts I had to make around the moldings on the closet and entrance doors, but I managed to make them as well.

Charlene and her mother were thrilled.

I had not billed Mrs. Brownlie for the installation, which would have been $1.25 per square yard or $15.

Mrs. Brownlie handed me a $20 bill.

"No, thanks," I said.

"But you didn't charge for the installation."

"I know. But I can't take any money from you."

"Why ever not? You did a fine job."

"I know. It does look like it came out okay. But you guys really did take a chance on me doing it, and I just don't feel I can take your money. I'm sure you can put it to much better use. Maybe it can help out with your birthday party." I looked at Charlene, who had a broad smile on her face as she looked at me. She really was a very cute and rather appealing young lady.

2

Why do these memories keep coming back to me? I have so many of them.

I now live in Massachusetts in a very nice home, with a lovely ocean view. We're in a wooded area; lots of wildlife come by from time to time. As I write this paragraph, it's April of 2021.

I got up this morning; fed our two cats; made some coffee; sat at our kitchen table, looking at the ocean in the distance; and turned on the TV, wondering what new horror would be in the news. Derek Chauvin had recently been convicted of murder—a White cop in Minneapolis, who the whole world witnessed choking George Floyd to death with his knee. Chauvin was White. Floyd was Black. So was Charlene.

Why did I think of her? Some fifty-eight years had gone by. My wife and I celebrated our sixtieth wedding anniversary last March. Why was I suddenly wondering about Mrs. Brownlie and Charlene? Mrs. Brownlie may be gone by now. But Charlene? Did she have her baby? Was it a boy or a girl? Did the father, who I never met, marry her, or did she raise that child alone with help from her mother? So much potential trouble and tragedy. I found myself fervently hoping Charlene and her baby, now grown, were okay. She was so nice and pretty and full of the loveliness that pregnancy can bring to a woman, when I knew her for those few hours.

But why did she and her mother and that linoleum come to me out of nowhere?

I suppose it found its way out of my subconsciousness; thanks to the turmoil we are going through today—not that different from what we were going through in Charlene's time. Martin Luther King and his movement were marching. Many Whites, including my father, supported him, marched with him. But the South was still loaded with Jim Crow. There was no way the Southern White governing bodies would ever give up segregation. In the North, we had the Black Panthers, Malcolm X. These were tough times for New Yorkers and most cities. The police were incredibly challenged. How to handle all the unrest?

One late afternoon, we were unloading a truck full of dirty rugs. The unloading area was on 157th Street at the back entrance of what used to be the movie theater. We pulled the rugs off the truck, carried them to a chute down, which the rugs fell to the basement, where the cleaning area was located. I was one of four guys pulling these rugs off.

Suddenly, some guy, who I had never seen before, came along the sidewalk and started yelling at me, "Yo, honky. Get outa our neighborhood." He went on and on at the top of his lungs.

I didn't, couldn't, say anything. I just stared at him. The guys I was working with didn't, couldn't either. I wasn't concerned for my safety. It never occurred to me. Sam, Eldred, and Ralph were with me. Finally, the guy ended his ranting and left. He might have noticed that his three "brothers" did not join in his attack.

As soon as the guy left, the three of them apologized to me profusely. "Bobby, we're so sorry!" I was Bobby to them. They had known me since I was quite young.

"It's not a problem," I said. And we continued unloading rugs.

Today, I doubt that guy would have taken off on me the way he did. Black Lives Matter, the George Floyd trial and conviction of Derek Chauvin for murder, and all the other Black deaths at the hands of police we have witnessed and continue to witness have brought on a countrywide, even worldwide, awareness of the racial prejudice issue in our country by a huge segment of the population, including all ethnicities.

It's a different time separated by several decades. Racial prejudice has never left us, but it has taken on a new color. We see actors of color on television we never would have seen in Charlene's time. We see Black politicians in places of importance we would not have seen back then. We recently had a Black president of the United States and, wonder of wonders, a woman of color as vice president. Things have changed. And yet parents of children of color still have to have "the talk" with their children when they begin to come of age. If a cop stops you, be sure and show both your hands to him. Don't be disrespectful. Do what the cop tells you, no matter what you think of it.

Did Mrs. Brownlie have "the talk" with Charlene? She must have. Yet she was so motherly toward me. I'll never forget her admonition: "You can do it." She convinced me I could install that linoleum. I never thought of that before, not until just now. Mrs. Brownlie was simply being a human being. The racial thing apparently wasn't considered. She probably knew that my father was not like…was not like what? How do you say it?

When my father was murdered in that very same store five years later, the local AME church held a memorial service for him. My mother and my wife, my cousin, my aunt, and I were the only White folks in the congregation. The church was filled to the rafters.

According to Reverend Harrison, Ellis Carpets was the last White-owned establishment in that part of Harlem. "He never gave up on us," said Reverend Harrison. "He never turned us down. If we had a need, he helped us. You would see him from time to time on his own hands and knees, installing some linoleum in a clinic we had just opened. He never charged us anything. He was a good man, and we honor him today."

*****

In September of 1972, at 6:00 p.m., Mildred had gone home, and my father had closed and locked the door behind her. My father was alone in the store. For some unexplained reason, the detectives theorized, my father had responded to a knock on the door and let two men in. He must have known one of them, said the detectives. He led them back to the office. He went into the office and spoke to these men through a window, at which people paid their bills, rather like a bank teller's window. It was dark. The fellow my father knew engaged my father in conversation. We have no idea what was said. Meantime, the other guy, who my father may not have known backed up into the shadows, worked his way around to the back of the office, where there was another door, found a heavy carpet kicker, opened this door, sneaked in, and struck my father over the head.

"He never knew what hit him," the detectives told me. They could tell by the way my father fell to the floor with his hands and arms at his side.

"Your father was almost deaf and going blind. He had no peripheral vision. So he didn't hear or see anything as the guy sneaked up on him."

"What do you mean he was going blind?" I asked the detective, who interviewed me later.

"The autopsy showed he had a huge brain tumor. He would have gone blind very soon."

3

I guess Mrs. Brownlie figured my father's son was the same. He wasn't, not really—but in her and Charlene's case, maybe so. Her apartment wasn't a clinic, and I charged them for the linoleum but not for the installation. I just couldn't.

Are you there, Charlene? Have you been marching in the Black Lives Matter protests, along with all those other people, all those Black folk, all those White folk, all those Asians filling the streets, carrying those signs? How about your child or your children, have they been with you? Did you, along with most of the rest of the world, celebrate the conviction of Chauvin?

Are you worried, though, about those that did not celebrate? Did you have the talk with your children?

Is my daughter going to have to have the talk with her son, my grandson? He's only four as I write. When does one have the talk with one's child? At what age? In their early teens, I suppose. So maybe ten years from now, will Carly have to sit down with Jackson and fill him in? Is there any chance that in ten years, it won't be necessary?

Jackson is a very cute, bright little guy, who is much loved by all his very White family. Well, we're not all pure White. His mother, Carly, suddenly discovered several years ago that she was a woman of color. One of her fellow teachers in the school where she teaches asked her if she would like to attend an annual conference held for people of color.

"I'm not a person of color."

"Yes, you are" was the answer.

"I am?"

"Yes, you are. Would you be interested in attending?"

"Uh, okay. Yes, I would," she said. She now attends these inspirational conferences annually.

Carly was born in Columbia, South America, and adopted by us, so she is a Latina. Her problem was that we, her parents, never treated her in any way other than as our daughter. The issue of skin color never came up, except that in the summer, we couldn't help but notice what a fantastic tan she got when she laid out at the beach. We often warned her to be sure and use sun tan lotion to protect herself against skin cancer.

Jackson, on the other hand, is a little boy of color. Carly adopted him four years ago. Will he need the talk?

My God, I hope not. I hope, somehow, we human beings can figure out that we are all members of a single species called *Homo sapiens*. One of the strangest things about this White supremacist phenomenon is the fact that *Homo sapiens* originated in Africa. If that doesn't make a joke of the so-called White supremacists, what does?

*****

A long time ago, when I was in grammar school at Public School 81 in Riverdale in the Bronx, our principal, Mrs. Simpson, came into our classroom. I was in the sixth or seventh grade. She announced to our class that tomorrow, several new children would be joining us, and we were to welcome them. That's all she said.

The next day, twelve new boys and girls came to PS 81. It seemed that a colored orphanage, which none of us knew existed in Riverdale, was being closed, and these were the last twelve children for whom no foster parents had been found. The kids were assigned seats in various classrooms. Ishmael Arrez was seated next to me. We became good friends. I brought him home to lunch a few times, having asked my mother if I could. My mother admitted to me many years later that she was startled when she first met Ishmael. I, of course, had not said or thought anything about the color of his skin. She welcomed him, never made a fuss. Why would she?

The best thing for us kids was that these new kids lived at the orphanage, which was a large very well-appointed facility, with a terrific playground. A whole bunch of us went over there from time to time to play.

All of us White and Black kids played together. We learned together in our classrooms. I don't believe any of the parents of us White kids had a problem. If they did, Mrs. Simpson must have somehow told them to shut up. I doubt if they did though.

I often wonder about Ishmael. Did he ever get placed in a foster home or, better yet, get adopted? What finally happened to that orphanage? Did all those kids get placed?

Another memory of early days: After I graduated from PS 81, I attended high school at McBurney School for boys in downtown Manhattan at 5 West 63rd Street just off Central Park West—the home of the YMCA at the same address.

Almost daily I am reminded of the cloakroom that was located at the entrance to the building. Members of the Y would check their coats on their way to whatever they intended to do.

The cloakroom was run by Joe. I don't remember Joe's last name. Joe was the only African American in the entire building, as far as I knew.

I have no recollection of how this came to be, but I got a job assisting Joe. I would receive coats, hang the coats on a large coat rack, and hand tags to the owners of the coats. But—and this is a big but for me—Joe taught me how to properly put a coat on a hanger before hanging it on the rack.

The trick is to grasp the coat by the inside collar with your hand, extend your elbow of the same arm into the coat, such that the coat's inside is fully opened, making inserting the hanger with your other hand very easy. It really works. And to this day, Joe's instruction comes to mind every time I hang anything in my closet. Every day, almost without fail, I think of Joe, my quiet, friendly coatroom instructor.

4

Recently, I attended a committee meeting for a nonprofit organization for which I volunteer.

Thanks to Black Lives Matter, the murder of George Floyd, and all the others during these past few years, we have concluded we need to create a statement which will reflect our organization's commitment to what is called Diversity, Equity, Inclusiveness (DEI). What an interesting time we have been having, trying to come to some kind of meaningful statement.

What an interesting time I have had examining my memories, my experiences. Who would have thought I would find myself questioning my own racial prejudices? Is some form of racial prejudice within me in spite of my own protestations to the contrary?

One of our members asked us to consider the following questions:

- When were you last treated with less courtesy?
- When were you treated with less respect?
- When and where did you receive poor service—i.e. in a restaurant, store?
- Do people act as though you are not as smart as they are?
- Do people act as though they are afraid of you?
- Do people act as though they think you are dishonest?
- Do people act as though they think they are better than you?
- Have you ever been called names or insulted?
- Have you ever been threatened or harassed?

All of us, senior citizens—all but one of whom were White—had no answer. We Whites could do no more than shake our heads.

Another early memory: Labor Day, Monday, September 5, 1949.

I had spent the weekend with my cousins in Leonia, New Jersey, just across the Hudson River from Manhattan, while my parents had decided to go away for the weekend. We lived in Yonkers, New York, just over the city line from New York City. My cousin John and I were fifteen years old.

While we were having lunch on Monday afternoon, the telephone rang. My uncle went into the living room to answer it. A few minutes later, he returned to the table. "That was your father, Bob," he said. "Seems he and your mother have had a rather bad time of it."

"Were they in an accident?" my aunt asked.

"I don't think so. It was something else. James said he'd tell us about it when they got here."

My parents arrived in Leonia in time for dinner. As they came into the living room, I asked them if they were okay as I hugged them.

"Well, yes. We're fine physically," Dad said. "But your mother and I have been through a horrendous experience. We were caught in a terrible riot. I wouldn't be surprised if people were killed."

"Good Lord. Where did this happen? What was it all about?" my uncle asked.

"In Peekskill."

"Peekskill?"

"Yes, Peekskill, New York. We decided to hitchhike and see where it might take us. We got a ride up to Ossining. Our driver let us off at a guesthouse not far from the Sing Sing prison. The next morning, Katherine saw an ad for a concert by Paul Robeson in the local newspaper. Another couple staying at the guesthouse was going to the concert and took us. The concert took place on a golf course with a large outdoor amphitheater-type area. Around twenty-five thousand people were there. It was a nice concert with a number of performers—Pete Seeger, Paul Robeson, and others I can't remember. We enjoyed ourselves.

"But at the end of the concert, it was announced that no one was to leave the concert area on foot.

"So the couple who brought us there took us and another couple in their car. It was a crowded car and a very hot day. But we were also admonished to not open our windows! Can you imagine that, not open our windows? *(Air-conditioning in cars did not exist.)*

"We were directed down a narrow dirt road, which was lined on both sides by thousands of American Legionnaires, Veterans of Foreign Wars, and state and local police. They were hurling rocks at us, swinging baseball bats at our windows, with the police just standing there and cheering them on! It was unbelievable. They were shouting at us, 'Hey, you, White niggers, go back to Russia.' Jew, Jew, go back to Jew town. We'll finish Hitler's job! Give us Robeson! We'll string that nigger up!' It went on and on."

Paul Robeson was a bass-baritone concert artist and stage and film actor, who became famous for his performances but also for his political activism. In the United States, he was active in the Civil Rights Movement and other social justice campaigns. His sympathies for the Soviet Union, where he said his skin color was ignored compared to what he called apartheid in America, caused him to be blacklisted.

The golf course was in Peekskill, New York, and the riot became known as the Peekskill riots.

I'll never forget my father describing this event, which, he later said, woke him up. We were in the height of the McCarthy Era when the so-called communist menace was at its height, and Senator Joseph McCarthy and other demigods were running rampant in our country. It was a very difficult time. Thousands of people were accused of being communist sympathizers and were blacklisted, thus losing their jobs. My father ended up joining the newly formed Westchester Committee for Human Rights and later formed the Yonkers Committee for Peace.

I have vivid memories of the mental turmoil this period in the history of the United States put me, my father and mother, and so many people through.

My father's battle continued while I went off to Swarthmore College just outside of Philadelphia. While I was there, he phoned me to say that the case *Ellis vs. Dixon et al. and the Board of Education of the City of Yonkers* was being heard the next day by the United States Supreme Court. Would I like to join him in Washington, DC? We met at the Baltimore and Ohio Railroad station in Philadelphia and rode together to Washington.

The Yonkers Committee for Peace had petitioned for the use of a hall in one of Yonkers' schools to hold a meeting. They had been denied. They had appealed through the New York State court system, including the New York Supreme Court. Now we were sitting in the courtroom of the United States Supreme Court. I remember how nervous my father's lawyer, Emmanuel Redfield of the American Civil Liberties Union, appeared as he pled Dad's case.

They lost the case five to four. But a few years later, after continuing the fight, Dad's committee won the same case, though regarding a later date for the meeting, and won in the Supreme Court of the State of New York. Things had changed. Tempers had calmed down.

5

A few years later, I graduated from college and made it through three years of law school. The military draft was still in effect during these years. I had a deferment. But the minute I graduated from law school at the University of Pennsylvania in 1959, I became eligible for the draft. I found a National Guard unit that could take me, and I joined them. Instead of a two-year commitment in the army, I was committed to six months active duty and five and a half years of weekly meetings in the Guard.

Soon after, I found myself going through basic training at Fort Dix in New Jersey. It's strange how I looked forward to entering the military with dread and how I now look back on my experience with some sense of satisfaction and enjoyment. One memory is of Sergeant Hamilton, who was the platoon leader of one of the four platoons in our company. We all occupied four barracks, which had been constructed during World War II. They had no heat and no hot water, unless we troops kept the coal-fired furnace going. It was August, so we mostly took cold showers.

The four sergeants who were in charge of our four platoons lived off base. Our sergeant in our barracks was Sergeant Grubb. What a name—not to be forgotten. He was a very nice guy, who kept us in shape and discipline without very much effort on his part as far as I can remember. Across the quad from our barracks, however, was Sergeant Hamilton's barracks. We, in our barracks, were so glad we were not in Sergeant Hamilton's barracks.

Every weekend, we had from noon on Saturday until Sunday night off. But that was not before we had inspection Saturday morning. So we cleaned the barracks thoroughly, cleaned our foot lockers, and placed every article we were supposed to have in perfect military order—toilet articles, shoes, socks rolled up correctly. We stood at attention in front of our double-decker beds in our dress uniforms, while Sergeant Grubb and a lieutenant none of us knew marched down the center aisle and looked us over. It was somewhat stressful. But we always passed.

So did Sergeant Hamilton's platoon, but not before he got his troops up bright and early on Saturday morning and made them remove their beds and footlockers from the barracks and line them up on the lawn in the quad, which was surrounded by our four barracks. Everything had to be in perfect order, duplicating the way everything was lined up inside. It was somewhat amusing for us, nonvictims of Sergeant Hamilton, watching his forty guys scurrying around, lining things up perfectly as Hamilton yelled at them. Then they had to clean the barracks and then bring everything back in making sure everything was spotless and then prepare for inspection. What a sight that was seeing all those beds and lockers lined up on that quad lawn every Saturday morning.

This was in 1959, and Sergeant Hamilton was the only African American in our company. I'll never forget him. He was, without question, the perfect soldier. His uniform was always crisply starched. He stood erect, with dark glasses and with several medals, which made it clear he had seen action and knew what he was doing. He commanded complete respect from everyone in our company.

On the day of our graduation from basic training, Sergeant Hamilton gave the commencement address. The entire company stood at attention in the quad as he complemented us on our accomplishments. It was the first time we saw a smile on his face. He admitted that to some extent, it had all been a kind of game that had turned us into soldiers.

I had not realized at that time what it had taken to make it possible for Sergeant Hamilton to be where he was. The United States military had been desegregated by Pres. Harry Truman eleven years earlier, in 1948, when he signed Executive Order 9981.

I learned later, however, that it was a decision made during the latter part of World War II, which led to Truman's decision.

Throughout World War II, Black American soldiers primarily served in labor operations in segregated units. Military leaders declared that Black soldiers were cowardly and had less inherent fighting ability. So they gave Black soldiers tasks, such as building bases, clearing mines, transporting troops and supplies, and grave registration.

Then in December 1944, during the last major battle of World War II called the Battle of the Bulge, the Allies were running low on manpower. It was proposed to General Eisenhower that the army could begin integrating Black soldiers with White infantry units. Black platoons would become full members of White infantry companies. General Patton, whose Third Army was crucial in the Battle of the Bulge, agreed to the proposal, and thus Black soldiers became eligible for frontline combat.

By March 1945, the US Army had crossed the Rhine River at Remagen, and the soldiers of K Company in the 394th Infantry, 99th Division, largely men from the segregationist South, had hunkered down under withering fire from the Germans. The advance of the Allies in this sector was stopped.

Then the K Company White soldiers saw Black troops advancing toward their position. After a fierce fight, K Company drove the Germans back, and the advance into Germany continued. According to one soldier from the South, who fought at Remagen, "The results were amazing. We ate together, slept together, fought together. There were no incidents." Soon, other divisions began integrating some of their regiments.

However, by the end of summer in 1945, the military was resegregated. Nevertheless, the success of those troops at the Battle of the Bulge helped Truman decide to integrate all the military branches in 1948.

When we look at the command structure of the US military today, we see a large portion of those who are leading our armed forces are African Americans, Asians, women. The success of integration in the US military turned the military into the most integrated element of our society from what was one of the most segregated.

But the military is as far removed from a democratic society as it possibly can be. The military has its own legal code. Leaders in the military are appointed, not elected. The president of the United States is the only nonmilitary citizen with authority over the military and all of its millions of members.

Nevertheless, I ask myself, could the military's success be duplicated in United States society?

As I look at what we all are going through now, I find myself wondering if things are different. The facts are certainly different. We're not struggling over the Russian communist menace anymore. Yes, we as a nation are dealing with the Russia that came out of the fall of the Soviet Union. But today's battle is different.

We have seen changes. Schools are no longer segregated, at least not legally. We no longer have colored and White bathrooms in the South. The green book, which was a travel guide African American travelers had to use to find lodging while wending their way through the South, no longer exists.

All the media, television, filmdom, and stage are multiracial. Most police departments today are multiracial and have women in uniform.

And yet various nonprofit and for-profit organizations find themselves having to affirm their commitment to DEI. And then, because of the COVID pandemic, Asian Americans found themselves under attack from their fellow American citizens, including African Americans.

Racism exists.

What can we do? What can I do?

I woke up, one morning, asking myself that question. *What can I do?*

Protest. Right. But I'm not of an age when I can spend hours in a crowd, wondering where the nearest toilet might be. So I'm not likely to join a physical protest in, say, Boston.

6

Sixty-three years ago, when I was first married, I could have protested—at least, in some manner, protested against what was happening in my own back yard that I knew was not right.

My in-laws were very nice, law-abiding folks, who owned a hotel. They had opened it right after World War II, in which my father-in-law had served for four years.

The hotel was a relatively small, thriving summer resort property. I helped out in the office and at the front desk from time to time. The hotel had a very regular seasonal clientele. People came for the season. Some stayed the entire summer, most for at least a month, very few for less than two weeks.

Here I was, the son of my extremely liberal father, now married into a very conservative family. I remember hearing my father-in-law saying that he supposed Senator McCarthy and the Peekskill rioters had gone a bit too far in their persecution of the commies. I believe he made this statement in the face of what his new son-in-law had brought to the conversational mix.

So be it. He had his point of view. My father had his. Both were decent men, who disagreed mightily. But they were civilized and always got along, at least on the surface.

Meantime, the hotel continued to thrive. One of the reasons it continued to do so, according to my in-laws, was the fact that they maintained their seasonal clientele year after year. If they looked at their reservation book in January and didn't see that they were very close to fully booked for the upcoming summer season, they got worried.

Who were their clientele? They were nice people—some of whom became close friends of my wife and myself, a few of whom attended our wedding. They also were as white as the driven snow.

We were admonished, in no uncertain terms, to not take reservations from Blacks. Period. Let any one of them in, and our regular clientele would disappear on us. It would produce an economic disaster for the hotel.

And I, the son of my father, went along with it. How do I explain that to myself?

My in-laws did business with a few vendors, who were Black. Their head cook in the kitchen was Black. My in-laws drove to Boston just before the beginning of the season each year, picked her up, and brought her out to the hotel, where she had a nice room with a little yard, where she grew things and lived reasonably well. We all loved her. At the end of the season, they drove her back to her home in Boston year after year.

In my defense, I will say that my wife and I took over the management of the hotel when my in-laws retired. Things had changed. People no longer stayed for long periods of time. They traveled from hotel to hotel during their vacations, doing the inns. Our average stay dropped from about three weeks to about three days! It was a very different touring world. Any restrictions regarding race or religion disappeared.

It was an economic decision—plain and simple—except I don't remember our ever having a discussion to the effect that we had to drop any policy regarding prejudice.

Our hotel was not unique in this respect. The Waldorf Astoria Hotel in New York City had the same policy during the first half of the twentieth century. It was so bad that when they headlined Black entertainers—Louis Armstrong, Marian Anderson, anyone Black—those entertainers could not enter the hotel through the front entrance. Rather they had to enter through a back entrance!

This was not in some Jim Crow Southern hotel. This was in New York City!

7

That was back in the first half of the twentieth century and beyond. Today we can say things have changed and for the better. There's no denying it. And yet, now in 2025, our country is unbelievably divided, not only politically, which we certainly are, but also racially, ethnically, religiously, culturally. There's no denying it.

Take a look at this photo:

What could possibly be wrong about these folks? In this age of theoretical enlightenment?

Recently, the organization *Americans United for the Separation of Church and State* (AU) was in federal appeals court, arguing on behalf of Gregg Tucker, who was fired for trying to address pervasive racism at the private school called *Faith Bible Chapel International* in Arvada, Colorado, where he worked as a teacher and director of student life.

After Gregg and his wife adopted a daughter, who is Black and from the Dominican Republic, some students began using racial slurs against Gregg and his family—actions that were part of a much larger and more pervasive problem in which White students dressed in Ku Klux Klan hoods, performed mock executions of students of color, posted racist White supremacist commentary on social media, and attacked their Black, Hispanic, and Asian peers with racial slurs.

Gregg tried to address racism by holding a symposium for students, which was backed by the administration. But when a small group of parents complained and demanded Gregg be fired, the school acquiesced. Gregg was stripped of his responsibilities as director of student life and was later fired.

Gregg sued the school. But to avoid responsibility for their actions, the school is trying to use religious freedom by invoking the legal premise of the ministerial exception as a defense for violating Gregg's rights.

What is the ministerial exception? It is an exception enunciated by the United States Supreme Court to the proscription in the United States Constitution, which prohibits Congress from making laws related to religion.

The First Amendment to the Constitution decrees, "Congress shall make no law respecting an establishment of religion, or prohibiting the free exercise thereof; or abridging the freedom of speech, or of the press; or the right of the people peaceably to assemble, and to petition the Government for a redress of grievances."

The ministerial exception is a legal doctrine in the United States, barring the application of antidiscrimination laws to religious institutions' *employment relationships with its ministers.* As explained by the supreme court, it seeks to both safeguard religious groups' "freedom…to select their own ministers," and to prevent "government involvement in ecclesiastical decisions." When applied, the exception operates to give religious institutions an affirmative defense when sued for discrimination by employees who qualify as ministers.

But which types of employees should qualify as a minister? And how broadly should the exception be applied. This was the subject of litigation regarding *Gregory Tucker v. Faith Bible Chapel International.* Gregg's attorneys argued that a dangerous trend has been created in which religious employers try to use religious freedom to license discrimination. They argued, "The ministerial exception was meant to ensure that houses of worship could freely choose their *clergy*. It was never intended to be a free pass for any religious employer to discriminate against its entire workforce and sidestep civil-rights laws." Gregg Tucker was never appointed as a minister, and his employment was strictly as a teacher, they say. As of this writing, the case has been appealed to the United States Supreme Court.

And now, one has to wonder what will happen when this case gets to the justices now on that court.

I can't help but roll my eyes when I think of this situation, where so-called Christian parents of students at a supposedly Christian school demand that the school fire one of its teachers because of his protestations against blatant racism openly displayed in the corridors of their institution.

What ever happened to Christianity? They claim to be Christians, followers of Jesus, who is quoted in the Bible as saying, "So in everything, do to others what you would have them do to you, for this sums up the Law and the Prophets" (Matthew 7:12).

I had a close friend, who was an avowed atheist who, every now and then, when commenting on something I had done, would say, "How very Christian of you." He always said it with a twinkle in his eye, but he also meant it as a compliment?

8

Humanity continues to be made up of a lot of very nice people doing nice things and of a lot of not-so-nice people, like a group of parents at the *Faith Bible Chapel International,* to say nothing about our current and former president and the current president of Russia, doing not-so-nice things, mean things to their fellow human beings.

Many of my former schoolmates and contemporaries—former judges, US senators, presidential candidates, people of great repute—are gone now. But here I am with a wonderful wife of more than sixty-four years, two very nice children who have very nice children of mixed racial backgrounds, two cats, and we have lots of adult friends of various religious persuasions.

As I go from day to day, I find myself at my rather advanced age, asking, *"Why? What is the point of it all?"* Does anyone else wonder? Does Charlene? Do her children? Does Joe? How about Sergeant Hamilton, Ishmael?

I suppose the answer to "*What is it all about?*" is, Well, we are here, all of us of all kinds of racial and ethnic diversities, all kinds of diverse—sometimes extremely diverse—beliefs and understandings, all of us, believers and nonbelievers, in everything from an amazing collection of religions,[6] collections of atheists, agnostics, believers in the rule of science, nonbelievers in the rule of science.

As long as we occupy this space on this planet, in its galaxy, in this universe, we might as well get along with one another. Why should I care whether or not you agree with me on anything? Go ahead. Believe what you want to believe, and I'll believe what I want to believe.

---

6 There are more than ten thousand religious sects on this planet.

I have a favorite question I ask myself every now and then.

Suppose our universe, in all its immensity, is conscious (which some cosmologists suggest). Suppose, as a conscious entity, it is not that different from ourselves. Within ourselves, all kinds of actions are taking place about which we know very little. Our blood flows, nourishing all kinds of cells, nourishing our brain, keeping things so that we can function with all our muscles, bones, lungs, heart, whatever. If you think about it, which I constantly do, our universe is not that different. It keeps going, expanding, planets revolving around stars, stars around galaxies, asteroids traveling here and there and all kinds of amazing, to us, other actions.

Suppose, then, the universe every now and then experiences various irritations, some very important, such as the explosion of a star, what we call a supernova, and some very minor, such as creatures on one of its rather small planets revolving around one of the smaller stars, having arguments with one another, maybe involving a good portion of the population on this planet, such as Putin's invasion of Ukraine, the Israeli, Hamas, Palestinian conflict. What would the universe think about that?

It probably would think very little about it, perhaps the way we think of a small itch we absent-mindedly scratch.

In the vastness, timeless, concern of the universe, this tiny planet's irritations would be of little, if any, consequence. If our universe could consult with a doctor, it probably would be told, "Don't worry about it. It will go away soon." Soon could be many of our years but insignificant in the context of the universe, which by the way, we are told is only one of possibly many universes. Some astrophysicists tell us that eventually our universe's expansion will cease, and then it will start to contract until it becomes an infinitesimal singularity, which will then start to expand again in a new big bang. Billions and billions of our years will go by during this recycling of our and all universes.

So why does all the pain and suffering and all the joy and happiness we experience on our planet Earth matter?

Why do I, do we *Homo sapiens*, concern ourselves over life as it exists on our planet? Why should I care about Charlene or George

Floyd's murder or my father's experience in Peekskill or his murder or whether or not US troops are desegregated or, more currently, the possibility of the United States becoming a dictatorship, which does have me greatly concerned?

9

It has been suggested to me that I should care because life is a gift and that I should cherish that gift.

According to Google, a gift is "a thing given willingly to someone without payment; a present, as in a Christmas gift."

Is life a thing? No? Then what is life? Google says, life is "the condition that distinguishes animal and plants from inorganic matter, including the capacity for growth, reproduction, functional activity, and continual change preceding death: existence, being, living, animation."

So one might say, life is a concept or a condition, which is identified in beings—beings being anything that grows, reproduces, continually changes prior to death? This to me suggests a thinking capacity of some kind, such as we see in our dogs and cats, in mice, in moles, bees, and insects. Some kind of process causes them to do what they do, no matter how mundane. And, of course, it suggests us humans with our so-called superior intelligence.

But this concept still begs a question. The definition of gift, as I have always understood it and as Google defines it, is something given. If it is given, then there must be a giver. So who or what gave *life* to all those creatures, big and small, and to all plant life?

According to the scientific community, life began in some kind of infinitesimal bacterial size in some kind of primordial soup on this Earth some four billion years ago. Once started, it was motivated to procreate, as are all living things, including humans. Procreate it did. And then various mutations came about. And very slowly but very surely, life began to be manifested in varying molecular sizes and then larger entities, until ultimately, all kinds of plant and animal life evolved—fish, mammal, reptile, plant, and insect. The process continued for many millions of years, until some kind of catastrophe occurred, which caused an extinction of a large portion of life as it was. But life had gotten hold on Earth and persisted, and a whole new evolution occurred, then was extinguished.

This process continued six times, and we are living in the sixth iteration of life on Earth.

There is a good size segment of humanity that chooses not to believe the science as described by evolution in spite of all the evidence, which has been discovered over the eons by us humans.

Rather, of course, most religionists would say, the answer is obvious. Life, even in its earliest stages, came from God. God created the universe, including all its solar systems, galaxies, asteroids, comets, black holes, planets, and of course, our Earth. And we can't say this for sure, but evidence seems to suggest that God not only started life on the Earth, but it is suggested that we are not alone, and life in some form probably exists on other planets throughout the universe, though we have no direct evidence to that effect. Astronomers, however, are finding other what they believe are planets revolving around other suns (stars), which, they speculate, could support life.

So let's say that Life is a gift from God, then I ask, Why are we doing everything we can to destroy that gift?

10

I am listening to some choral music at this moment—gorgeous stuff. Every bit of it is filled with what my atheistic friend would have called the Christian myth, so much drivel. But what beautiful drivel—brings me to tears. I have sung Fauré's *Requiem* so many times as a member of Chorus North Shore or the Cape Ann Symphony Chorus or other organizations. Why my tears? My atheistic friend sang it with me in Chorus North Shore, and he had tears in his eyes.

Fauré's *In Paradisum* from his Requiem in D minor, Op. 48—incredibly beautiful.

On the other hand, I have buried several family members. Recently, I saw my cousin-in-law's casket in the ground, where it had rested for quite a few years as we buried her husband, my cousin, next to her in a similar stainless-steel container with the casket inside. Her body, and now my cousin's, will be well-preserved for many decades, if not centuries. To what end? Where is this paradise Fauré describes? This, I have to confess, was not a pleasant experience.

The problem with getting as old as I am is that I have so many memories—some troubling, some sad, embarrassing. They keep coming to mind. Sometimes they make it hard to go to sleep.

One which keeps coming to me for some reason was rather amusing at the time, I suppose.

I was probably around six years old. We lived in a two-family house with another family, the Higensons, upstairs from us. They had a child, who probably was around three or four. His name was Billy. Billy's parents worked during the day, and they had a maid come in every day to care for Billy. One day, I was upstairs for some reason and chatting with the maid, who was Black. She was a very nice lady, very friendly. I remember noticing that the palms of her hands were of a very light complexion. In my innocence, I asked her if she was turning White. She laughed out loud and called to Billy. "Billy, Bobby thinks I'm losing my chocolate!"

Another memory from around the same time comes to mind. There was a coal-fired furnace in the basement, which heated the house. Periodically, coal had to be delivered. Our two-family home was situated on a slight hill so that our basement was above the street level. Normally, coal was delivered via a chute, which sent the coal from the truck to the coal bin in one's basement. But in our case, that wouldn't work. The window, overlooking the coal bin, was too high. So a workman had to carry the coal to the bin. He had a large sturdy bag, which he slung over his shoulder. He pulled a lever on the truck, and the coal poured out into the bag. Once filled, he carried the coal up the slight hill to a window in the basement and poured the coal into the bin.

What I remember so vividly from about some eighty years ago was that the load of coal was obviously very heavy. But the truck driver, who was White, didn't help this man at all. He just sat in his truck, smoking a cigarette, while the workman walked back and forth many times, carrying that load. He was an African American. At that time, which was around 1940 or '41, he was called a nigger.

I remember feeling so sorry for him. I'm having that same feeling now as I write. Why? What is going on with me? Why does it matter? It was so long ago, eighty or more years. I have no idea who he was, where he lived. Was he married? Did he have children? He was alive. He was a fellow human being.

I suppose all of these memories matter because I am a human being and humans, unlike most other mammals, have brains, which somehow accommodate what we call feelings. I heard a zoologist recently who said that the creatures she was taking care of live in the now. They don't think about the future or the past, just what is important to them in their now, like hunger or, in the wild, the threat of losing their life by being eaten by another creature. But we know that some mammalian creatures do seem to think a bit more widely. Elephants have been seen mourning the death of a fellow elephant. Some so-called higher primates have shown some basic emotions. In a recent experiment, monkeys were observed not choosing to pull a chain for their food because they knew pulling that chain caused a neighboring monkey to experience pain via an electric shock.

Here's my big "however!"

I am, at this moment, living a relatively comfortable life with my wife of sixty-four years. My son and his wife and family live next door. My daughter lives a fair distance away but not so far that we don't see her and her now seven-year-old son fairly frequently. Our immediate surroundings are very pleasant, being surrounded by woodlands and with the Atlantic Ocean in view. Recently, we arrived home from a recital we had participated in, and I had to admit to myself, as I drove along our rather lengthy driveway and saw my somewhat contemporary home, that my seventeen-year-old granddaughter had it right.

A short while ago, all of us were returning from a walk through the woods on various paths by the several former granite quarries that are filled with water. She was by my side as she suddenly exclaimed, "We are so lucky to live here!" Yes, we are. Our town is a tourist Mecca. People spend thousands of dollars to visit our area for a week or two.

But we live here. My ninety-one years have brought me here, to this moment.

So why am I not content?

Not that long ago, I explained to a Roman Catholic priest, who is a friend of mine, the philosophy or tenets of Christian Science. I had been born, raised, and lived the first half of my adult life following its tenets. By the time I was chatting with my friend, I had left Christian Science.[7] As I finished my explanation, he asked me, "But where is there reason for hope?"

Christian Science takes the teachings and practices of Jesus very seriously, so much so that it teaches that there was a science behind

[7] My experiences as a Christian Scientist and the path which led to my leaving is explained in my memoir *A Collision of Truths, A Life in Conflict with a Cherished Faith*, https://www.amazon.com/s?k=Acollision+of+truths&i=stripbooks&crid=319S7SMDSITBL&sprefix=acollision+of+truths%2Cstripbooks%2C81&ref=nb_sb_noss.

Jesus's miracles, which we can learn and from which we can experience healing. But I don't recall any suggestion of the location of any place called the kingdom of God. In fact, in Christian Science services Jesus is often quoted as saying, "The kingdom of God is within you."[8]

The sense of hope that my friend was wondering about, I believe, came from a belief that upon dying, we humans would travel in some way to this kingdom of God called heaven. Faure's *paradisum?*

My atheistic friend called this wishful thinking.

I do not have this hope. When I die, I will be cremated. There will be no attempt to preserve my body for some kind of miraculous awakening. The Bible makes it clear. The English Standard Version of the Bible declares, "By the sweat of your face you shall eat bread, till you return to the ground, for out of it you were taken; for you are dust, and to dust you shall return."[9]

Of course the New Testament brings us a rather different perspective, preaching a heaven to which all nice people will go when they die and a hell to which all not nice people will go, unless they have somehow been granted absolution from their sins by another mortal to whom they confess their sins or unless they manage to come to Jesus by some method described by the particular congregation to which they belong, before it's too late.

And probably some 90 percent of my friends and acquaintances adhere to this belief in one way or another.

And I love them.

---

[8] Luke 17:20–21: "The Kingdom of God does not come visibly, nor will people say, 'Here it is,' or 'There it is,' because the Kingdom of God is within you."

[9] Genesis 3, 19.

11

And so I keep wondering what's going to happen in the near future and more distantly.

Within some time, in the not-so-distant future, my life will end. My physical body will be cremated—whoosh…gone up in smoke. Some ashes will be disposed of by whomever—my children—in some manner determined by them.

And what will the future bring for them and for my grandchildren and their children and their children?

No matter what amount of ranting and raving various *Homo sapiens* engage in, no matter what inability of said *Homo sapiens* to get along with one another, no matter how many insurrections, riots, whatever, it won't matter a hill of beans.

It has been pretty well established that the human species is going to disappear—go extinct. Possibly other life forms—vegetable, various mammalian, aquatic, insect, whatever—may survive.

However, projections are being expressed that a superior "intelligence" will replace our intelligence. And this superior intelligence will have, ironically, been created by our intelligence.

Well then, we may ask, will this intelligence be a life form?

Billions of dollars are being spent in various laboratories all over the world, developing algorithms which can learn. Many relatively crude learning algorithms exist right now. You've experienced their learning powers. They make predictions regarding how you may react to various stimulations, such as your buying habits.

Much more sophisticated algorithms are being developed, however, which will learn how to react to various stimuli—good or bad—regarding our human existence in medicine and eventually in all walks of life.

In his book, *The Kraken Project*, Douglas Preston[10] describes a NASA project in which a vehicle, called *Explorer*, is to be launched for the purpose of exploring Saturn's moon, Titan. At the heart of this vehicle is its computer programming, which will allow it to travel many years before finally arriving at Titan. Its problem will be that the distance is so great that, should *Explorer* run into a problem, it will not be able to communicate with those on Earth, who are responsible for its well-being in time to receive a solution. Even at the speed of light, it will take too long for us earthlings to receive a message and respond and alter its course of action before it is too late. So *Explorer* must be able to react on its own. It must learn how to react to all possible stimuli, be they positive or negative.

Hence, an artificial intelligence (AI) algorithm of great intelligence has been created and installed in Titan *Explorer*. Included in the package is an amazing array of devices and tools, which *Explorer* can use to react to and overcome various obstacles, which it might encounter.

Preston's *The Kraken Project* is a work of fiction. But just how far-fetched are the problems which *Explorer's* scientists encounter? It's a very good read, so I won't give too much away. But one problem they encounter is that, in a test of *Explorer's* capabilities to react to various threatening stimuli, things go awry. And *Explorer*'s "intelligence" (or survival algorithm) runs amok. In short, in his book, Preston describes troubling possible developments in so-called artificial intelligence.

Remember *Star Wars*' C-3P0?

C-3P0 knew six million languages! He/it also had a kind of *Homo sapiens* appearance. It probably was originally designed by humans, who, of course, thought theirs was the ultimate intelli-

[10] https://www.amazon.com/Kraken-Project-Wyman-Ford/dp/0765317699.

gent design. But in the last analysis, it was a machine—a machine designed by humans and given super AI. Six million languages? With its AI, C-3P0 could undoubtedly solve problems much more quickly and efficiently than us humans.

What is logically wrong with the following? Mankind develops robots with AI of a most advanced nature. Said robots have no need for consuming any animal or plant life. Their skin is covered in solar cells. All their energy comes from the sun.

Initially, they are programmed to value all life. They see humanity doing everything it can to destroy said life. So why not see humans as a kind of weed in the garden of Earth. Get rid of them.

Like us humans, machines do wear out over time. Our C-3P0s will need to be maintained and, eventually, replaced. Hence, C-3P0s will need to be created to meet these needs—repairing broken circuit boards, solar panels, also to create the various components of individual C-3P0s, metals, precious (to humans) elements needed for a C-3P0's computer/brain. The C-3P0 species, however, will be totally classless. Those needed for mining precious metals will be designed for that purpose. Those needed for assembling new C-3P0s will be designed for that purpose. The list goes on and on.

So why does the Earth need us humans?

Then, of course, the question must be asked: Why does the Earth need C-3P0s?

Once humanity has gone extinct?

Ultimately, this planet we call Earth is going to disappear, as are all planets in this solar system of ours and, eventually, all planets in the universe. In the meantime, however, other solar systems will come into existence, maybe including Earths, with so-called life.

This last prognosis made by today's cosmologists won't occur for several billions of years, so we have no need to become fretful. No? Take a look at this graph.

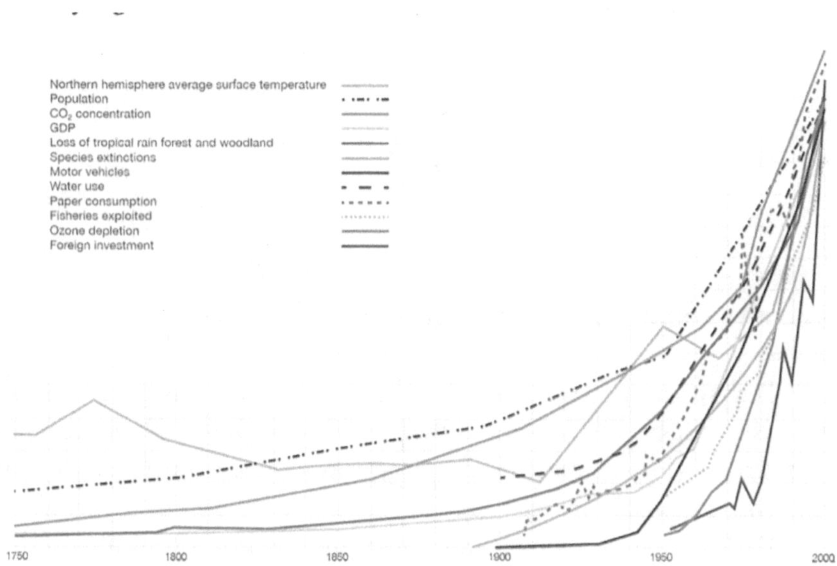

The above graph, published by the World Health organization (WHO)[11], shows, "Key environmental issues deemed by the WHO to have the greatest impact on global health. The list includes: Demand for clean water, global surface temperatures, ozone depletion, consumption of ocean resources, species extinction, CO_2 concentration, deforestation, and global sea levels.

"All of these negative indicators have been on the rise over the last century. Now," according to the WHO, "they are all accelerating at terrifying rates."

[11] Taken from Dan Brown's book *Inferno.*

12

When we come right down to the nitty gritty, what is our existence all about?

Perhaps more to the point, What has human existence been all about since it began? We are known as *Homo sapiens. Homo* refers to manlike two-legged creatures, and *sapiens* comes from the Latin for "one who knows."

Early on, there were several humanlike creatures walking on this Earth—Neanderthal, *Homo erectus*, to name a couple. Slowly but surely, all but *Homo sapiens* were extinguished, Neanderthal being the most recent extinction.

Why did these other creatures disappear? Because they could not survive the conditions they faced. As I noted earlier, there is some speculation that the Neanderthals were driven into nonexistence by us *Homo sapiens*, however, not before there was a certain amount of intercourse between the two, leaving some Neanderthal blood flowing in the veins of us *Homo sapiens*.

It would seem, then, that it was our superior intelligence that made it possible for *Homo sapiens* to win the two-legged mankind race.

But just think of all the incredible pain and suffering that accompanied this win. How many individuals, of whatever hominid, suffered pain and suffering at the hands of their fellow hominids? Hundreds of thousands of years of suffering.

We don't have to look that far back to see such cruelty. Today, right here in the United States of America, we are witnessing attempts by various groups of *Homo sapiens* to reign terror over their neighbors, who they consider to be undesirable. Think of the so-called White nationalists and the (supposedly) Christian right.

Even worse, as I write, President Putin of Russia is attempting to annihilate thousands of the inhabitants of Ukraine. And prime minister Netanyahu of Israeli has killed thousands of inhabitants of Gaza. A ceasefire is supposedly in the works as I write. But who knows whether it will work and whether or not another attempt at killing off a bunch of our fellow *Homo sapiens* will come about?

But let's set aside the *Homo sapiens*-directed cruelty and look at the unfortunate members of our species, who simply are down and out. This is one area where we *sapiens* can claim some solace. There are thousands of both governmental and nongovernmental organizations which are attempting to address the evils facing us *sapiens* today and bring some level of comfort to the suffering. This, of course, is as it should be.

But I have to ask, Will it never end? These various genocides, holocausts, and all the poverty that exists worldwide have always been with us. For as long as we *Homo sapiens* have trod this Earth, these evils have gone on.

All to what end? *What is the point of our hominid existence?*

My question, of course, assumes that some kind of initial design started it all, had a meaningful purpose, and that it all went awry—really awry, starting at the beginning, when initial life forms started immediately consuming other life forms for the purpose of surviving and being able to procreate. The process continues to this day.

Is there no point to it all? None?

Should we say that everything just is, all that munching by us of our fellow animals—cows, pigs, fish, et al.—all the munching of their fellow creatures and plants, all their production of methane gas, all of hominids' destruction of the very living thing upon which we live—all of it? All of the pain and suffering and angst-producing activity simply *is*?

A very close friend of mine, who died a few years ago of old age, knew, in his bones, that there had to be a reason for existence (life), there had to be some kind of first cause that brought it all on. He was an atheist, and he was a deep thinker who believed that there was something dreadfully wrong about quantum theory and its conclusions that everything ultimately is brought on by chance. Renowned quantum theorist, Werner Heisenberg, argued that reality did not have independent inherent form, rather it was created by the process of observation. Experiments in quantum theory have demonstrated that it is the observation by the physicist in his/her lab that brings about the result.

One of my friend's arguments was to ask, "What caused the big bang?" The beginning of it all, that great explosion that astrophysicists proclaim, brought on our universe. Before the big bang, they say, a singularity existed. Smaller than a subatomic particle, this singularity contained all that was needed for what is today. Some fourteen billion years ago, it ended its singular existence and began to expand at an unimaginable rate, thus letting loose an evolution of everything, including life, which began about four billion years ago. Our universe, they say, is still expanding at what some physicists say is an alarming rate.

My friend wondered if the big bang theory was valid. If it is, then he asked, "What brought about the singularity's existence and, therefore, our universe?"

There had to be some kind of force or some kind of organizing entity to design and then start the process. But then, one must wonder, What brought on that organizing entity? And whatever brought the previous entity on and on and on and on? And one must ask, Where was this singularity located? Assuming all the countless universes, which are speculated on within the scientific community, are part of the result of the big bang, how did the singularity exist?

He was/is not alone in this wonderment.

13

Of course we view all these questions within a time frame. Possibly the most important time frame for us is the duration of our life. We know when we were born, in many cases, down to the actual date and time, and we constantly speculate about when our existence on this Earth will end. What date? How many years will we have accumulated?

But our sense of time depends entirely on where our Earth exists within our universe. Our year equals one rotation of our Earth around our sun. Our day equals one rotation of our Earth around its axis. We know that our Earth makes a single rotation on its axis 365 days during a single rotation around our sun, except that every four years, we have to add a day to the year, giving us 366 days. So our system is not perfect. Over the millennia of hominid existence on this planet, many different calendars have been invented to help hominids figure out their particular when.

Our universe, as a whole, has no such concern. Even our estimation that the big bang occurred fourteen billion years ago, that our Earth was created some four billion years ago, that our Earth will be consumed by our sun when it becomes a supernova in some four billion years is a *Homo sapiens* invention based only on our observation of the relationship between our Earth and our sun. These calculations on our part are of no consequence when it comes to the universe. The universe *is*. All its movements and changes *are*, period. Our sense of time does not exist within the calculus of the universe, nor within all the hundreds, if not thousands, of universes speculated upon by cosmologists. In fact, it is even speculated that those physical laws cosmologists rely upon so vigorously are unique to our universe, that other universes we speculate about may have entirely different and unique to them physical laws, which control their existence.[12]

When astronauts travel outside of the Earth's atmosphere, we tend to think of them as traveling through nothingness, a vacuum. If their space suit were to develop a leak, their oxygen would escape into that vacuum. The astronaut, too, would be sucked into the void we call space if his space suit didn't prevent it.

But physicists discovered in the late nineteenth and early twentieth centuries that space has properties independent of the existence or nonexistence of matter. And when physicists, including Einstein, thought about it, they realized that time and space must be considered together when considering all the bodies in the universe. They call it space-time. They even discovered that space-time causes light to bend around large bodies. When observing an eclipse of the sun, they saw that a star's light bent around our sun on its way to the observers on Earth.

All this is neither here nor there, except to say that we, on this Earth, have no clue what's going on in our universe, although the folks at NASA and other such institutions say that they do, sort of.

I have tried and tried to grasp their reality. The most I have come to understand is that I don't really understand it except to say that it forces me to question the validity of the reality that I observe on this Earth to say nothing of all the thousands, if not billions, of realities each of us humans believe to be correct.

*****

It has been suggested to me that I am a nihilist. Am I? Do I believe that life is meaningless? Do I reject all religious and moral principles?

I believe I can safely say that I do not reject all moral principles. The golden rule works just fine for me.

Religious principles? Hmm. How many wars of annihilation have been fought and, indeed, are being fought at this instant, thanks

[12] See https://en.wikipedia.org/wiki/The_Accidental_Universe.

to religious so-called principles. Think of the Tucker family and their battle with their so-called Christian school. Look at today's White Christian nationalists, who are battling physical and legal battles right now. Indeed, how many wars of annihilation have been inflicted on *Homo sapiens* and, before them, all *Homo* species over the millennia? It goes on and on and appears to bring never-ending suffering.

Add to that, all life exists by consuming life.

What kind of meaningfulness to life, as we know it, does any of this suggest? What does the future portend, and bottom line, what is the point of it all?

14

I already am a partially bionic *Homo sapiens*. Perhaps you are too. I have two fake knees. They are metallic and require that I inform TSA attendants at airports that they will sound an alarm.

Do you have an artificial implant of some kind, maybe a new hip? Maybe a prosthetic that allows you to do something you would not otherwise be able to do?

As mentioned earlier, in 2005, inventor and futurist Ray Kurzweil wrote *The Singularity Is Near: When Humans Transcend Biology* in which he projected a future where artificial intelligence would become supremely important, and he suggested that implants for the brain would be developed, which would supersede any powers of calculations we humans now have.

Then, in 2016, Yuval Harari published *Homo Deus: A Brief History of Tomorrow* in which he predicted developments in medicine and AI, which will bring us to the point where we will *no longer be Homo sapiens*, rather we will become a new species, which he suggests might be called *Homo Deus. Homo* means man. *Deus* means god. This may seem rather fanciful, but his suggestion is that AI and all the physical and medical developments that will come out of it will make it possible for a new species to evolve, which will be able to live forever, not immortal. Accidents will occur, causing the death of individuals.

His suggestion also says that we will see these startling developments within the next few decades. Hello, C-3P0?

Harari states that he is not prophesying when he describes these developments; rather he is suggesting the future he describes is perfectly possible and that we must be aware of these potential developments and be wise in the choices we make.

Indeed, we must. What would a future in which a species became able to live forever mean for the overpopulation of the Earth, which presently is reaching alarming numbers?

I have to admit that as I read *Homo Deus*, I became totally involved in the pages. But at the same time, I found myself getting a bit depressed, or perhaps rather sad would be a better way to put it.

What about the creative process? Consider all the art that has been produced over the eons. And how about music?

If a new species evolves, such as *Homo Deus*, will said species have an appreciation of what I can only call feelings or emotions, appreciation of the intangibles, the nonmaterial, products of what we call the mind? While the new species may become supremely intelligent when it comes to calculations regarding the mechanics of our bodies, will present day love of music continue? Will my appreciation of Fauré's *In Paradisum*, in which the mythical religious heaven filled with celestial beings is described, be understandable to this evolved species? While I do not believe the myth, I still am greatly moved by the music. It evokes emotions of I do not know what. All I know is they are there. That goes for so much music and so many artistic expressions we, in this day, encounter.

What is the mind? That is where all these conditions occur, isn't it? Is it somehow lodged in the brain? The brain, when held in one's hand, is merely a gelatinous, rather strange-looking goop. Some folks eat brain. When that brain was in an animal's head, it had all sorts of electrical impulses, causing all kinds of connections that, at least to us humans, appear as thoughts and feelings. Are all our thoughts and feelings merely electrical connections?

Okay then, I come back to my original question: Why? Why do I care? What is the point of it all? Coming back to our place in the universe or universes, why does any of this matter?

I see no reason why our existence matters in the universal sense. The universe and all its components, all its suns, planets, asteroids, moons, and on and on are performing all kinds of physical, cosmological actions and will continue to do so, possibly forever. In fact, it has to be forever. Otherwise, there would be a void, a nothingness, which I cannot fathom.

Therefore, is there any meaningful reason for my/our existence on our Earth? I can't think of any. Can you?

Does that make me a nihilist? I care about the well-being of my family, my pets, my friends, and neighbors. I care about what is happening, as I write, in Ukraine and Gaza. I worry over what kind of world my grandchildren are going to face. I concern myself over animal welfare, wild and domestic, separation of church and state, forces working to destroy our democracy. I donate to many different organizations concerned with these issues. I also volunteer for the Cape Ann Symphony, an excellent regional symphony. Why do I do these things? I guess it is because they bring some sort of sense of meaning to my own personal existence.

But has anything actually changed in the history of *Homo sapiens*, with all the scientific progress, all of the diversity, equity, and inclusion (DEI) efforts, et al? Are we *Homo sapiens* even capable of changing our tendencies? Is it built into our DNA?

During the reign of Queen Elizabeth, when Catholics were murdering Huguenots in the thousands, and Huguenots were murdering Catholics in the thousands, a sixteenth century poet wrote the following:

> Tomorrow, and tomorrow, and tomorrow,
> Creeps in this petty pace from day to day,
> To the last syllable of recorded time;
> And all our yesterdays have lighted fools
> The way to dusty death. Out, out, brief candle!
> Life's but a walking shadow, a poor player,
> That struts and frets his hour upon the stage,
> And then is heard no more. It is a tale
> Told by an idiot, full of sound and fury,
> Signifying nothing.[13]

Shakespeare's Lady MacBeth was in a very depressed state of mind when she uttered those words. But was she wrong?

---

[13] Macbeth, Act V, Scene 5, William Shakespeare (1564–1616).

Predictions are being put forth by responsible people that our democracy may be in danger. In fact, it is suggested by Harari that, ultimately, his suggested new species, which will supersede *Homo sapiens*, will have no room for a democratic government.

Why does the *Homo sapiens* species exist? Why are we here? Given all the incredible angst and pain, we *Homo sapiens* have caused and experienced, over the thousands of years of our existence, what earthly or universal purpose can there be to my/our existence.

My dictionary says *purpose* means "the reason something is done or created." And *reason* means "a cause, explanation, or justification for an action or event."

Put these two together, and what do you get? The definition of purpose could be limited to any kind of causal relationship, as in physics or cosmology. No thinking entity is required to make Einstein's $E = mc^2$ valid. But *reason*, especially when you accept the word *justification*, does suggest a thinking causal source—some kind of consciousness.

Another way of putting the question could be, What earthly or universal *meaning* can there be to my/our existence?

My thesaurus defines *meaning* as "value, validity, worth, consequence, account, usefulness, use, significance, point."

Some sort of thinking or consciousness is needed to tell us what the point is of my/our/Earth's/the universe's existence.

What might that consciousness be?

The answer coming to most people's minds has to be God, Allah, Jehovah.

*****

All the above description and speculation on my part has resulted from decades of studying all sorts of treatises, mostly regarding the cosmological/quantum theory conundrums, which exist today. But I was brought up and lived the first half of my life as a Christian Scientist. It was all the above reading, which caused me to leave the church.

But the evolution of my thought did not cause me to dismiss the whole of Christian Science. I wonder if anyone who has been brought up within a particular belief system ever loses the whole of their upbringing.

Christian Science was founded by Mary Baker Eddy in 1865, right after the civil war. This was a time when the medical profession was very crude. During the war, doctors cut off a soldier's leg using a saw, which had not been cleaned after the amputation of another soldier's leg.

This also was a time when all kinds of charlatans, called quacks, were selling various so-called healing methods.

Christian Science, which offered an alternative to then current medical procedures and quackery, became a panacea for millions and became a large religious movement.

From my perspective, however, the medical question was not paramount. My father was not a Christian Scientist, and he and my mother brought me to the doctor from time to time. As a young lad, I had my tonsils removed. I remember a doctor calling on our home to check me out and confirm that I had measles. I don't recall any treatment being prescribed. However, my mother, for sure, prayed mightily.

As I progressed into adulthood, I began to see that Eddy did not only call for reliance on God for healing; she also contemplated the wonders of the sciences as they were developing. I imagine she was aware of Albert Einstein. She died in 1910. He put forth his theory of relativity in 1905, which caused an incredible reaction within the scientific community and the world at large.

One of the quotes from Eddy's writings was "What can fathom infinity?" not who. *What* can fathom. To my mind, she was seeing something way beyond the problems of healing the human body.

She put forth seven synonyms for God. Every Christian Scientist knows these synonyms. I do, to this day, some forty-five years after resigning from the church. They are life, truth, principle, spirit, mind, love, soul. In her writings, of which there were many, she often spoke of divine mind.

Divine mind! Hmm. I suspect, this synonym for God made it possible for me to contemplate what a number of scientists have wondered: *Is the universe conscious?*[14]

Quantum theorists tell us that for things to occur in their lab, the scientist's *observation* causes the result. This says to me that for any result to come about in the world of atoms and subatomic particles, some kind of observation is taking place constantly.

For me, the question then becomes, Is this a purposeful observation?

Some sort of thinking or consciousness is needed to tell us what the point is of my/our/the Earth's/the universe's existence.

Isn't it?

What might that consciousness be?

How would the atheist respond? According to my dictionary, an atheist is a person who disbelieves in a god or gods. Interestingly, there are many cosmologists, physicists, scientists who also are devoted Christians of many denominations, or Jews or Muslims who, nevertheless, pursue these questions. So what causal entity (consciousness) are they/we searching for? My answer is, "I do not know." Amit Goswami, cited in footnote 14, says the universe is "self-aware." My thesaurus says, to be aware, one must be conscious.

Right now, as I write, we are doing everything we humans can conceive of to ensure our extinction as a species—multiplying ourselves at an alarming rate; slowly but surely clearing all forests, thereby threatening our supply of breathable air; polluting what breathable air we still have left; heating up said air, thus changing our climates worldwide; killing off wildlife. The list seems to be never ending.

Some of us are trying to reverse these trends. But can we reverse them in time to save ourselves?

I can't help asking once again, *What is the point of it all?* Our planet keeps spinning on its axis. Our moon revolves around us. They both keep revolving around our star, along with the other planets in our solar system. Our sun keeps producing the energy we feel on our

---

[14] See *The Self-Aware Universe: How Consciousness Creates the Material World* by physicist and cosmologist Amit Goswami, PhD.

faces on a sunny day. And all this action will continue for another four billion years. If *Homo sapiens* were to cease to exist, then slowly but surely, new life and surviving plant and animal life would come along, and a proper balance of life would evolve.

Or would it? We *Homo sapiens* are not only working very hard at ruining our existence on this planet of ours; we also are working hard at replacing ourselves with a new species of artificial, nonbiological, artificially intelligent entity.

In the grander universal scheme of things, one has to ask why? What earthly or universal benefit would be achieved? We, supposedly intelligent living creatures, have brains, which have evolved to the point where we wonder about things. Hence all our discoveries about how our bodies work and can be fixed. Hence all our discoveries of other locations on Earth folks knew nothing about. Hence all our explorations of outer space using all kinds of machines we have invented for that purpose.

All to what end? If we go extinct and leave this Earth to a population of C-3P0s, what will be their purpose? Universally speaking, I can think of no purpose. Can you?

And what about the arts? We *Homo sapiens* (theoretically *wise men*) have, over the centuries, managed to create an amazing array of works of art musically, visually, architecturally, sculpturally. My wife and I belonged to three choruses and have helped several other choral organizations. We have participated in various plays. We are very active with the Cape Ann Symphony, a top-notch local organization, which entertains many communities north of Boston. Our children and grandchildren are wonderfully creative.

The arts in all their forms are very important to us. My wife and I have toured Europe and Africa, participating in choral tours and marveling at the thousands of works of art and architectural wonders on offer. Most of the choral works we sing are products of mental creations about a world, which does not exist. Nevertheless, as in Fauré's *In Paradisum*, they are so incredibly beautiful.

Will a non-*Homo sapiens* species, a C-3P0 species, have the capacity or even the need to appreciate such, to us, artistic wonders—wonders which bring tears to our eyes? Will a C-3P0 have an artificially intelligent bunch of chips, which can produce tearful reactions to the sounds we react to?

I don't know. But I do know that we *Homo sapiens* are an incredibly selfish bunch, who care little about all these issues, most especially climate change and overpopulation of *Homo sapiens*, us humans.

If we do not pay attention, starting yesterday, to the worldwide population explosion, we will be doomed. We, men/women, who are supposedly wise—Homo sapiens—will in the not-that-distant future disappear from this Earth. And our sun, solar system, galaxy, universe won't give a damn.

15

In his bestselling book *UFO: The Inside Story of the US Government's Search for Alien Life Here—and Out There*, Garrett M. Graff[15] notes that there are millions of solar systems in our universe. Is it not reasonable, he asks, to assume there are, therefore, millions of planets out there, at least some of which are life bearing?

Let's try a mental experiment.

Let us assume that among all these possibly life-bearing planets, there is a planet somewhere that is supporting life. For the sake of this piece, let's call it Earth2. Let's also assume that said life is intelligent, so intelligent that some of its members wonder if there is life on some other planet.

We, on our planet, have no way of knowing of the existence of this planet, but our unknown Earth2 residents have solved the limitation of the speed of light. They are able to go into what *Star Wars* called "warp speed." Possibly worse, they have discovered a planet many hundreds of millions of miles away, which shows evidence of having some kind of life—our Earth.

Their planet is somewhat larger than our Earth and a bit farther away from their star, which is a bit larger than our sun. Hence their days and years last longer than ours. And so it goes.

Let's further assume that scientists on Earth2 have also found that life, in many forms, does exist on our Earth! Indeed, one of the life forms is rather intelligent, so intelligent it might be something we, on Earth2, should investigate.

However, folks on Earth2 have been observing our Earth for some time—several centuries. And they have seen that the highest

[15] https://en.wikipedia.org/wiki/Garrett_Graff.

form of intelligent creatures have a propensity for killing one another, sometimes killing entire populations in one fell swoop. While their many inventions have evolved rather interestingly over our Earth's centuries to the point where they are pointing various telescopic machines into space in a seeming attempt to understand the cosmos, they, nevertheless, keep killing one another.

It has also been observed that these seemingly intelligent creatures, in spite of their propensity to kill one another, have multiplied at an amazing rate in a very short period of Earth's time to such an extent that the very existence of these creatures is threatened. And they have suddenly increased the average atmospheric temperatures of their Earth at such an alarming rate to literally assure their very own extinction!

After much discussion on Earth2, it is not surprisingly decided that no matter how curious this Earth may be, we on Earth2 had better leave it alone. We on Earth2 have managed to exist in reasonable harmony with one another, no matter whatever it is that causes Earth1's populations to function the way they do. There are many other possibly life-supporting planets in our galaxy without threatening to infect ourselves with whatever it is that has infected Earth1. Let's leave this one alone.

Infected Earth1! Is there some kind of sickness that exists in our DNA that assures the ultimate destruction of our species? While other mammalian species on our planet do eat one aother, they do so for physical sustenance, not for the fun of it. Interestingly, there is only one species that does kill other life forms for the thrill of the kill.

16

I have to make a confession. I have fond memories of that time during basic training in the army when I had possession of an M1 rifle and of firing it during target practice. The M1 was the primary weapon for us troops. My M1 weighed eleven pounds, as I recall. When we were hiking to bivouac with our full packs on our backs, those eleven pounds meant something. You knew you were carrying a real weapon.

The M1 was a semiautomatic weapon with eight rounds in a clip. The M1 installed each round into the firing chamber automatically, but I had to pull the trigger for it to fire that round after which the next round was chambered automatically. One of the memories I have is of being on the firing range, along with a bunch of other guys, each with their M1, and firing at large targets that were one thousand yards away. That's right, a little more than half a mile away. That's how powerful that weapon was. You really could aim at and hit that target accurately from that distance.

But what was most important was that when you fired more than one round, the next rounds should hit very close to the first. When you fired several rounds, they all should end up in a tight shot group. I'll never forget when I was sprawled out, aiming my M1 and pulling that trigger for each shot, a sergeant I did not realize was watching from behind me called out, "That's mighty fine shootin' soldier." I hadn't quite hit the bull's-eye, but my shot group was very close and tight.

Why do I bring this up? Because I enjoyed firing that M1 at target practice while at Fort Dix, New Jersey. And my confession is that I completely understand a person enjoying firing a weapon at a target. I also can understand folks having a fascination for and collecting these things. They are complicated and interesting devices.

However, I abhor the concept of firing a weapon at any living creature. I could never try to kill a deer crossing our lawn or a bear or coyote or fox in our woods, not even a squirrel of which there are plenty running around outside right now. I could not go hunting.

I am not suggesting that hunting is wrong. Our forebears survived by hunting for their next meal. Today, many hunters seek sustenance in the same way as their forebears. But I do not believe legitimate hunters use AR-15s or whatever other type of multishot automatic. The idea is to kill the creature in as humane a way as possible, bring it back, and prepare its meat for consumption. Sometimes deer or other creatures get hit by a car. Generally speaking, state or local police will save the carcass and give it to a needy family.

This, as far as my knowledge goes, is as it has always been, given that we are *Homo sapiens*, residing on this Earth. Of course, in the early days, we did not have guns. We had bows and arrows; before that, spears and clubs.

Eventually, though, weapons evolved to include guns. Prior to the American Revolution, during and after, weaponry had evolved to include the single shot front-loaded musket. In battle, an experienced soldier could fire and reload as many as three times a minute, meaning the firing rate was at most three balls (not bullets) in one minute. Today the AR-15 or 16 or whatever can fire twenty, even hundreds, of rounds in a single burst in a few seconds.

Perhaps you can see where I am going with this.

We have, as I write, a pandemic going in the United States, which is not related to a biological disease. It is killing thousands of US citizens a year, including many children. A large portion of these deaths occur in schools throughout the United States.

What is the cause of these deaths? The cause is the distribution of hundreds of rounds of ammunition dispersed in seconds, not a musket-fired three balls a minute.

It was the musket that Thomas Jefferson, John Adams, et al. had in mind when they wrote the Second Amendment to the United States Constitution, which reads, "A well-regulated Militia, being necessary to the security of a free State, the right of the people to keep and bear Arms, shall not be infringed."

There is no way back in 1791 that those men, who drafted the ten amendments to the constitution, envisaged the evolution of weaponry to the assault weapons of today. There is no way, back in 1791, those drafters pictured a United States of America having a population of some 332,000,000 *Homo sapiens.*

Think of it. The population of the USA was about a quarter the size of England's and a sixth the size of France's, only about 3,930,000 (about 1 percent of today's population). An overwhelmingly rural population lived in the new America, about half in the northern states and half in the South.

Out of this population, various military units were formed by the colonies and then, after the Revolutionary War, the states—the minutemen being the most well-known and best trained. It was these men, with their muskets, who fought the British men with their muskets. It was these men who formed the well-regulated militia of the thirteen colonies turned states. It was these men who maintained their muskets in their homes, as did most households at the ready in case they were needed to defend their home or for the purpose of finding their next dinner. It is out of this tradition that the Second Amendment was created.

In those early days, children went to school in little one-room schoolhouses. No member of society in 1791 ever dreamed of bringing their three-shot-a-minute musket to one of those schoolhouses and firing their musket at those innocent children and their teacher.

It was unthinkable then. It is unthinkable today.

So what is our situation today?

Contrary to all claims to the contrary, proliferation of assault weapons in homes in the United States was never approved by the United States Supreme Court. The gun lobby has repeatedly sought to invalidate lifesaving gun safety laws. But the United States Supreme Court has consistently recognized that the Second Amendment is compatible with *strong firearm regulations.*

In 1994, Congress passed the Assault Weapons Ban. The bill passed with a sunset provision of a decade in place, meaning that when lawmakers agreed to it, they knew that it would automatically expire in 2004, unless renewed through another vote. Expire it did. But it has not been overturned by any court.

Then, in 2008, The District of Columbia (Washington DC), passed a law, which prohibited the ownership of handguns in the home. This law was challenged and came before the supreme court. In the case of District of Columbia v. Heller, the court overruled the district's prohibition of ownership of handguns in the home. But Justice Antonin Scalia wrote, "Like most rights, the right secured by the Second Amendment is not unlimited. *[It is] not a right to keep and carry any weapon whatsoever in any manner whatsoever and for whatever purpose*" (Italics added.).

In addition, the court spelled out prohibitions, which would be presumptively lawful, including the following:

- Prohibit firearm possession by dangerous people.
- Forbid firearm possession in sensitive places, such as schools and government buildings.
- Impose conditions on the commercial sale of firearms.

Then two years later, in 2010, in *McDonald v. City of Chicago*, the supreme court held that the Second Amendment applies to state and local governments, as well as to the federal government, which, in Heller, only applied to the District of Columbia and, therefore, the federal government.

While Chicago's complete handgun ban was overturned, the court reiterated that a wide variety of state and local gun laws are constitutionally permissible. Importantly, the McDonald court quoted *Heller*, saying: "It is important to keep in mind that *Heller*, while striking down a law that prohibited the possession of handguns in the home, recognized that the right to keep and bear arms is not 'a right to keep and carry any weapon whatsoever in any manner whatsoever and for whatever purpose'" (quoting Justice Scalia).

And yet, despite all this adjudication, we have a gun pandemic on our hands.

Will it never end? I suspect that it will. It probably will require a few decades of wrangling by us *Homo sapiens* in the United States of America to come up with a meaningful firearms control system.

But to what end? And how many fellow *Homo sapiens* will need to be slaughtered before a solution is finally agreed upon?

Of course, while I was writing the above paragraphs, several more mass shootings occurred.

Why do I care? Why do I keep asking myself, *"What's it all about?"*

Why do I keep going back in my mind to Lady Macbeth's lines:

> Life's but a walking shadow, a poor player,
> That struts and frets his hour upon the stage,
> And then is heard no more. It is a tale
> Told by an idiot, full of sound and fury,
> Signifying nothing.

"*Because*," I keep saying to myself, "*we Homo sapiens, year after year, decade after decade, century after century, millennia after millennia, keep proving Lady Macbeth's (Shakespeare's) depressing lament to be correct.*"

Look at what we have come to! We once had a rather steady population of *Homo sapiens* of a few hundred million worldwide. Today it is eight billion. We are slowly but surely crowding out all other life on this planet. How many species have gone extinct because of loss of their habitat; thanks to human activity.

Meantime, thanks to all our seemingly miraculous scientific discoveries, we are, slowly but surely assuring the end of the *Homo sapiens* species—us. Some kind of new species is going to replace us. But this new species will be enhanced with all kinds of mechanical devices loaded with artificial intelligence, which will eliminate the need for large parts of the human brain or will replace the human brain entirely. Perhaps this new species will be entirely mechanical, not alive in the way we define life, not biological—a machine, a robot.

What will be the purpose of this species?

17

As noted in the prologue, women in the United States are fighting fiercely to overcome the rampant onslaught by a bunch of mostly old men in high places, along with Biblical fanatics and so-called Christian nationalists, which appear to include justices of the United States Supreme Court, aimed at denying the right to have control of their own health care, especially abortion rights. And as if that's not enough, in the US, we have the ongoing battle over the right to bear arms and, while we are at it, killing a bunch of people on a daily basis. Then there are the more silent issues. The constant onslaught on the constitution's guarantee of separation of church and state, and all sorts of attacks on our rights as citizens to exist without governmental interference. All kinds of organizations exist for the purpose of defending these principles—the American Civil Liberties Union, Americans United for the Separation of Church and State, to name only two.

Will these conflicts never end? Will they ever be resolved?

If we look at the history of the world, the answer has to be no. Possibly, current conflicts, disagreements, battles for superiority, whatever will be resolved. But at what cost in human life? How many years will these kinds of conflicts last, and how many humans will suffer or be killed in the process. And then what? As present-day issues get resolved, you can be sure, new ones will arise.

I suggested earlier that it is hopeless, that it is in our collective DNA. How many megalomaniacs can we list who are infecting our way of life right now? Worldwide, there are thousands of *Homo sapiens* in government and the business worldwide who thrive on their megalomaniacal lives.

And, yes, there are thousands of us, plain old human beings, who care and are trying to fight the megalomaniacal challenge.

And then there are the literally billions of us *Homo sapiens* who are lost, swimming in drugs, poverty, sickness without aid. There are organizations trying to reach out to and assist these people, but it seems to be hopeless.

We are at a crossroads—all of us, eight billion *Homo sapiens*, including the worst of us and the best of us.

We *Homo sapiens* have been on this Earth for many millennia and, no matter which time we pick to have a look at what existence must have been like, no matter how wondrous their buildings, their tombs, we find conflict. Nations have risen and fallen in a never-ending process right up to the present. How many nations have disappeared in the last one hundred years? And one has to ask, How many millions of souls suffered inquisitorial tortures, gas oven murders, incredible lashings at the slave owner's whips, woundings and deaths at the end of an AK-16 and continue, to this day, to this hour, even to this minute.

Some kind of hugely major change has to occur among us, eight billion, and soon. Our Earth won't notice or care. Whatever life on our Earth survives the present and following eons won't matter one bit to the Earth, its sun, its solar system, our galaxy, the universe.

There's no doubt about this. For the next thousands, millions, even billions of years, what we do for ourselves today and in our ensuing years will be of no consequence. Unless…

It is entirely up to us—us *Homo sapiens*—to figure out how to survive!

Are we up to the challenge?

Will modern day scientific developments save the day?

Can modern day scientific developments save the day?

Will we *Homo sapiens* save the day?

Can we *Homo sapiens* save the day?

*****

It seems that every time I think I've run out of things to wonder about, something new to me comes about. Shortly before he died, I saw Henry Kissinger in an interview. He was one hundred years old. In the interview, he brought up a concern about the future I had not thought about.

His concern? Artificial intelligence. It's coming faster and faster. There's no avoiding it. But that was not his concern. It is inevitable that it will evolve, and we will see it in all walks of life before we know it. But his question was, "Who will develop it to the point where international diplomacy will rely upon it? What diplomatic principles will it reflect? Will our (USA) diplomatic principles be paramount? Or China's? Or Russia's? One might also ask what will our principles look like?

Is it inevitable that all the many conflicting national goals will continue, no matter the extinction of our species—wars upon wars, megalomania upon megalomania, ego upon ego? Will we slowly but surely go extinct? Will—can—a new species evolve with improved thinking abilities, thanks to AI, which will in some way change our DNA and rid us of megalomania?

If the answer in some way is yes, my question still is, *why*? What is the point? Why does this Earth, this solar system, galaxy, universe need a *Homo* species?

18

I have grave concerns over the possibility that my grandchildren and their children will face some incredible challenges.

I am sitting at my desk, typing these words but also peering out my window from time to time. It is early May 2025. We are experiencing some showers, which will bring May flowers. I see leaves beginning to pop out of the bare branches of forsythia. Yellow blossoms are in abundance. A squirrel just ran across the deck. A deer went by, followed by her very cute young fawn. Seagulls fly by. Various birds are singing their mating songs and building their nests for their eggs. We have several crows who populate our trees with their loud raucous calls.

Recently, a neighbor sent me these pictures from a nearby state park, and it all makes me feel a sense of sadness.

I am so very grateful for my present situation. I am blessed.

But right now, the future poses an incredibly unpleasant picture.

Can mankind, *Homo sapiens*, do anything about it?

No matter how you look at it, everything we have done to ourselves and what we see happening as the future comes and goes, it seems to be a hopeless picture, at least for us *Homo sapiens*!

But it may be even worse!

Recently, I was sitting on a tree stump in the woods and found myself fascinated by a tiny bug that was working its way across the leaves at my feet. I had no idea what kind of bug it was, nor what its mission might be. But it was a rather cute little thing as it walked over my boot toe, somewhat like a brown lady bug, and I found myself wishing it well.

Why do I mention this? Because I recently read the following from *Skeptical Inquirer*:[16]

> Insects can seem to be everywhere, all at once—sometimes to an annoying degree. After all, three out of every four known animal species on Earth are insects. But these dazzlingly adept creatures—which pre-date the dinosaurs—are suffering a silent and hugely consequential crisis, with their numbers plummeting around the world.

We are doing everything we can right now to destroy the existence of the honey bee, who we love to see pollinating plant life. With all our pesticides, we also are doing everything we can to, ultimately, kill off insects of all kinds. Life depends on the existence of these creatures. They literally create the healthy Earth all plant life needs. Without all those sometimes yucky-looking worms and other creepy crawlies, dirt cannot feed plant life.

Is all life on our planet ultimately doomed,—not just us eight billion *Homo sapiens*, which we suspect are already doomed to go extinct sometime in the not-too-distant geological future but all life? Thanks to the incredible growth in the population of us on this Earth and all our ingenious inventions and chemical creations, we have brought on what has to be the greatest threat literally to all life!

I just heard/saw a commentary on MSNBC that some politicians are becoming concerned over the possible consequences of AI.

---

[16] https://skepticalinquirer.org/.

Really? Only now? The concern appears to be that we *Homo sapiens* are slowly but surely assuring our extinction. My, oh, my! What a revelation!

So what is it all about?

I have had to come to the conclusion that despite the good intentions of so many folks trying to alleviate the pains and fears of so many humans and wild and plant life worldwide, unless some kind of transformative revelation occurs among the vast majority of us *Homo sapiens*, we are doomed. I fear for my grandchildren's and most especially their children's future.

At my age, I probably need not be terribly concerned. The woodlands that surround us, the deer, raccoons, squirrels, chipmunks, wild turkeys, and others that come by from time to time probably will continue to do so.

My wife and I will continue to enjoy the love and company of our son and daughter and their children and so many friends of all kinds of religions and beliefs, which give one a sense of hope.

As I write, I'm ninety-one years old.

How many more years do I have? Who knows?

But here's what I do know; We *Homo Sapiens* face an incredible global crisis. According to the publication *Free Inquiry*, there are some ten thousand religious sects worldwide! Can all our religious beliefs, nonreligious beliefs, haters of science, lovers of science, agnostics, atheists, physicists, chemists, cosmologists *ever come together and meet the challenge?*

Epilogue

I attended a family reunion in Albuquerque, New Mexico, a short while ago.

The reunion was brought about by a memorial service for my last cousin, Mary Lou. I had three Ellis cousins, and she was the last. Her children, my other two cousin's children, my son, and several grandchildren were there.

I am the last in our family of my generation, so it was wonderfully meaningful for me. Also, I knew nothing but bare bones about Mary Lou since we all went our separate ways starting with college. I learned a lot about all my relatives and their accomplishments. There were a ton of hugs to go with it all. All the gang ended up asking me many questions about various mysteries they wondered about, concerning the early days of the Ellis family before they were born.

I did not want to go. I was concerned about leaving my wife alone as she was having serious balance issues, and there was no way I was going to travel alone. But my son, Jeff, told me I had to go, and he made sure I didn't get lost along the way (most of the folks there were of his generation). His son and daughter said they would take care of their Nana, and they did.

Why do I describe this episode?

Because I found myself wondering how many other folks have similar get-togethers.

If you have read *Too Much and Never Enough*[17] by Donald Trump's niece, Mary L Trump, or *All in the Family: The Trumps and*

[17] https://en.wikipedia.org/wiki/Too_Much_and_Never_Enough and https://www.simonandschuster.com/books/All-in-the-Family/Fred-C-Trump/9781668072172

How We Got This Way, by Donald's nephew, Fred Trump, you know that not all families, no matter how wealthy or prominent, have friendly get-togethers, if any at all. The monumental tensions that still exist in the Trump family may never be healed.

It is saddening to me to see people, some of whom I know, have difficult family issues.

If megalomania, narcism (which seem to be a large factor in such relationships) is prevalent in many human relationships. How can we ever hope to heal this *Homo sapiens* problem? How can we ever hope to survive as a species?

*****

All of what I have written in these pages is, in the last analysis, speculation.

Is our species bound to succumb to a sixth extinction?

Are the *Homo Sapiens* species bound to become AI creatures?

Earlier I quoted Shakespeare's *MacBeth*. Now I quote Shakespeare in his *Hamlet*, in which Hamlet declares the following:

> To be, or not to be, that is the question:
> Whether 'tis nobler in the mind to suffer
> The slings and arrows of outrageous fortune,
> Or to take arms against a sea of troubles,
> And by opposing end them: to die, to sleep
> No more; and by a sleep, to say we end
> The heart-ache, and the thousand natural shocks
> That Flesh is heir to? 'Tis a consummation
> Devoutly to be wished. To die, to sleep,
> To sleep, perchance to Dream; aye, there's the rub,
> For in that sleep of death, what dreams may come,
> When we have shuffled off this mortal coil,
> Must give us pause. There's the respect
> That makes Calamity of so long life:
> For who would bear the Whips and Scorns of time,

The Oppressor's wrong, the proud man's contumely,
The pangs of despised Love, the law's delay,
The insolence of office, and the spurns
That patient merit of th'unworthy takes,
When he himself might his Quietus make
With a bare Bodkin? Who would burdens bear,
grunt and sweat under a weary life,
But that the dread of something after death,
The undiscovered country, from whose bourn
No traveller returns, puzzles the will,
And makes us rather bear those ills we have,
Than fly to others that we know not of?
Thus conscience does make cowards of us all,
And thus the native hue of Resolution
Is sicklied o'er, with the pale cast of Thought,
And enterprises of great pith and moment,
With this regard their Currents turn away,
And lose the name of Action.

Shakespeare wrote these words in 1599–1600. Has nothing but mechanical and medicinal evolution occurred since then?

We, all of us, in the last analysis, are distant cousins. Think about it. All of us, no matter the color of our skin, the language we speak, where on this Earth we reside are distant cousins. We all are members of the eight-billion-member family called *Homo sapiens.*

Is the extinction of us *Homo sapiens* inevitable? Will it matter what kind of species, if anything, follows?

Does our planet care?

Does our solar system care?

Does the universe care?

Does any member of the Homo sapiens species care?

Do you care?

About the Author

Robert Y. Ellis holds a degree in economics from Swarthmore College and a law degree from the University of Pennsylvania. Among his several careers, he was a journalist with the Christian Science Monitor, Executive Secretary of the United States Figure Skating Association, editor of *Skating Magazine* and is author of *A Collision of Truths* and *An Act of God*. Currently, he is a member of two choral organizations and serves on the board of directors of the Cape Ann Symphony. He and his wife reside along the rockbound coast of Massachusetts.

www.ingramcontent.com/pod-product-compliance
Lightning Source LLC
LaVergne TN
LVHW090428210926
848480LV00002B/729